Intersex Matters

SUNY series in Queer Politics and Cultures

Cynthia Burack and Jyl J. Josephson, editors

Intersex Matters

Biomedical Embodiment, Gender Regulation, and Transnational Activism

DAVID A. RUBIN

Published by State University of New York Press, Albany

Printed in the United States of America

For information, contact State University of New York Press, Albany, NY
www.sunypress.edu

Production, Diane Ganeles
Marketing, Michael Campochiaro

Library of Congress Cataloging-in-Publication Data

Names: Rubin, David A., 1978– author.
Title: Intersex matters : biomedical embodiment, gender regulation, and transnational activism / David A. Rubin.
Description: Albany, NY : State University of New York Press, [2017] | Series: SUNY series in queer politics and cultures | Includes bibliographical references and index.
Identifiers: LCCN 2016059171 (print) | LCCN 2017020893 (ebook) | ISBN 9781438467566 (ebook) | ISBN 9781438467559 (hardcover : alk. paper)
Subjects: LCSH: Intersex people—Identity. | Intersexuality. | Gender identity.
Classification: LCC HQ78 (ebook) | LCC HQ78 .R83 2017 (print) | DDC 306.76/85—dc23
LC record available at https://lccn.loc.gov/2016059171

10 9 8 7 6 5 4 3 2 1

For my parents,
Terrill Kay Eliseuson and Jerome Leon Rubin

And for Max Beck and Berky Abreu, in Memoriam

Contents

Acknowledgments

This book owes its livelihood to friends, family, advisors, and colleagues who generously gave their time, feedback, advice, and encouragement along the way. I am profoundly indebted to the individuals and communities who helped to make this project a reality.

The book saw its beginnings nearly a decade ago in my dissertation project for the Women's Studies department at Emory University. I am grateful to my chair, Lynne Huffer, and my committee, Rosemarie Garland-Thomson and Holloway Sparks, for taking me under their wings, commenting on countless drafts, and sharing their wisdom when I needed it most. Lynne has been and remains a mentor without parallel. For her invaluable advice and steadfast support, I cannot thank Lynne enough. Thanks are due as well to my master's thesis advisor, Miranda Joseph, for her remarkable generosity and kindness, and to Sandra Soto, Charlie Bertsch, and Laura Briggs for their formative inspiration and support. I thank my undergraduate teachers Sue Ellen Jacobs, Tani E. Barlow, Priti Ramamurthy, Aimee Carrillo Rowe, Shirley J. Yee, Steven J. Shapiro, and Matthew Sparke for nurturing my interest in feminist inquiry. Phil Tobin and Stacy Yee, two other early mentors, deserve special thanks for their abiding support and warmth.

Most of this book was written and rewritten while I was teaching in the Women's and Gender Studies program at Vanderbilt University and at my current institutional home, the Department of Women's and Gender Studies at the University of South Florida (USF). At Vanderbilt, thanks are due to Charlotte Pierce-Baker, Houston A. Baker, Ellen Armour, Katharine Crawford, Rory Dicker, Julie Fesmire, and the one and only Barbara R. Kaeser. I am tremendously grateful to my colleagues in Women's and Gender Studies at USF—Diane Price Herndl, Elizabeth Bell, Kim Golombisky, Michelle Hughes Miller, Milton Wendland, Jessie Turner, Rondrea Mathis, and Jennifer Ellerman-Queen—for welcoming me into the department with open arms and supporting my work. It is a true honor to be a member of such a generative and lively academic community. Diane deserves extra-special thanks for her guidance and mentorship since my arrival at USF.

Research for Chapter 5 was supported by a 2014 summer Research Grant from the Humanities Institute at USF. Thank you to Elizabeth Bird, Elizabeth Kicak, and staff of the Humanities Institute for their assistance.

For helping me to think through many of the ideas in this book, I thank the members of my "Politics of Women's Health," "Transnational Feminisms," and "Queer Feminist Science Studies" undergraduate and graduate seminars at USF. In particular, I thank Sandra Carpenter, Laura Leisinger, Ella Browning, Azure Samuels, Mary Dickman, Kelly Dyer, Sunahtah Jones, Richard Henry, Kaajal Patel, Paola Rivera, Jennie Reiken, Viki Peer, and Mary McKelvie.

For generously sharing insights about intersex, medicalization, and activism, I am indebted to Max Beck, David Cameron Strachan, Marcus Arana aka Tio, Anne Tamar-Mattis, Caitlin Petrakis Childs, Lynell Stephani Long, Iain Morland, Georgiann Davis, Morgan Holmes, Hilary Malatino, Monica Casper, and Irie Keiko.

Thank you as well to friends, colleagues, teachers, and editors who helped me to grow over the years: Berky Abreu, Mark Jordan, Jack Halberstam, Agatha Beins, T. Denean Sharpley-Whiting, Rez Pullen, Elizabeth Vennel, Carla Freeman, Beth Reingold, Elizabeth Wilson, Martine Brownley, Cynthia Willett, Geoffrey Bennington, Eric Hayot, Claire Nouvet, Cathy Caruth, Shoshana Felman, Deepika Bahri, Jonathan Goldberg, Michael Moon, Gayatri Gopinath, Carla Freccero, Banu Subramaniem, Larin R. McLaughlin, Moya Bailey, Stefanie Speanburg, Jae Turner, Sarah Prince, Brandi Simula, Shruthi Vissa, Shannan Palma, Linda Calloway, Elizabeth Kennedy, Monique Wittig, Suresh Somnath Raval, Paul C. Taylor, Amy Harrington, LeKeisha Hughes, Lee Burkey, Karen Leong, Roberta Chevrette, Cricket Keating, Clint Harris, Samantha Ruehlman, Kyle Lighthiser, Erin Durben-Albrecht, Sara Giordano, Deboleena Roy, Courtney Berger, Robyn Wiegman, Sara Jo Cohen, Meridith Kruse, Alberto Delgado, Susan Stryker, Richard Morrison, Antionio Viego, Ron Dobson, Roderick Ferguson, Mary Hawkesworth, Miranda Outman, Andrew Mazzaschi, and Michelle Richards.

Huge thanks to Stanley Thangaraj for being a cherished friend and trusted ally. Thanks to Angela Willey, Kristina Gupta, and Cyd Cipolla, my co-conspirators and collaborators in queer feminist science studies, for indispensible advice, commiseration, and camaraderie. Thanks to Zach Zulauf, Scott Lindsay, Ayako Takamori, Rehka Kuver, and Juliet Ceballos for their long-standing care and support.

I am very lucky to have a consummate philosopher-companion in Solan Jascha Jensen, who I thank for always being by my side, even when he is halfway around the world. Words cannot adequately express how much this work owes its existence to my dear friend Joshua Marie Wilkinson. For his unshakeable faith, sage advice at literally every turn, abiding critical questions throughout good times and bad, and unparalleled loyalty, I am beyond thankful to Joshua.

For their unconditional love since day one, I thank my family: Terrill Kay Eliseuson, Jerome Leon Rubin, Alex Rubin, my aunts, uncles, and cousins, and the late Mildred Rubin. I especially want to thank my parents for their sustained material and emotional support and for imparting in me their belief in the value of critical thinking. Tony deserves special recognition for always reminding me of what's important. Endless thanks to Erica Brundidge for much-needed laughter, joy, and sustenance. The struggle continues.

Beth Bouloukos has been a kind and dedicated editor. Thanks to Beth, Rafael Chaiken, Diane Ganeles, Michael Campochiaro, and the entire team at SUNY for being incredibly helpful throughout this process. I am grateful to David Luljak for his careful assistance in compiling the index. Thank you to the anonymous readers of this manuscript for their extremely generous and perceptive feedback, criticisms, and suggestions for revision. Any errors, flaws, or omissions that remain are, of course, entirely my own.

Parts of this book have appeared elsewhere in different forms. An earlier version of Chapter 1 was published as "'An Unnamed Blank that Craved a Name': A Genealogy of Intersex as Gender," in *Signs: Journal of Women in Culture and Society* (2012). An earlier version of Chapter 5 and part of the conclusion appeared in "Provincializing Intersex: U.S. Intersex Activism, Human Rights, and Transnational Body Politics," in *Frontiers: A Journal of Women's Studies* (2015). I am grateful to University of Chicago Press and University of Nebraska Press for permission to republish the contents of these articles.

Introduction

Intersex Matters

> There are humans . . . who live and breathe in the interstices of this binary relation, showing that it is not exhaustive; it is not necessary.
>
> —Judith Butler[1]

Intersex is an umbrella term for the myriad characteristics of people born with sexual anatomies that various societies deem to be nonstandard. I say "people born with sexual anatomies that various societies *deem* to be nonstandard" rather than "people born with nonstandard sexual anatomies" to call attention to the material-semiotic overdetermination of intersex.[2] The prefix *inter-* literally means "among, between, in the midst of." But it would not be quite right to say that people with intersex anatomies exist "between the sexes," nor that all those diagnosed or self-identified as intersex live and breathe in what Judith Butler calls "the interstices" of the "binary relation" of gender, though some surely do, just as many non-intersex people do.[3] Rather, it is at the levels of biological and cultural intelligibility, broadly construed, that intersex embodiment disrupts binary schemas of sex and gender. For this reason, intersex has not been accepted as part of the order of things. Since the mid-twentieth century, western biomedicine has managed these interstitial bodily figures through surgery.

While intersex generally refers to people born with anatomies that defy received understandings of the nature of sexual difference, within and beyond the medical community the meaning of intersex is contested. The genetic, chromosomal, and biochemical etiologies or causes of atypical sex development are heterogeneous and complex.[4] Medical professionals are not required to record the number of individuals with intersex conditions whom they treat. Furthermore, intersex people are not included in mainstream demographic surveys or census bureau data. For these and other reasons,

the frequency of intersex is notoriously difficult to estimate.[5] Nonetheless, a commonly cited figure holds that approximately 1 to 2 infants per every 2,000 are physiologically classifiable as intersex.[6] Of those, many, though not all, are subject to surgical and/or hormonal normalization in the metropolitan global north, and recent research and activism makes evident that this trend extends into some parts of the global south as well.[7]

According to the prevailing biomedical point of view, individuals with intersex conditions, or what have recently been renamed as disorders of sex development (DSD), are born with "abnormalities" that require a medical fix. While physicians use surgery as a corrective measure, surgical normalization frequently entails problematic consequences for patients.[8] For instance, surgery often leads to patients' partial or total loss of capacity for genital sensation, as well as various recurrent health problems.[9] According to ethnographies of twentieth- and twenty-first-century intersex individuals' experiences of treatment by Sharon Preves, Katrina Karkazis, and Georgiann Davis, as well as numerous testimonies by intersex persons, activists, and their allies, infant genital surgeries cause considerable physical and psychological pain and distress for affected parties.[10] In addition, the claim that infant genital surgeries lead to problem-free outcomes lacks quantitative and qualitative support.[11] Most importantly, perhaps, the vast majority of intersex variations—there are many, and they are more common than one might think—present few health risks, and in most cases no more health risks than those with which non-intersex individuals are born.[12] These factors have led a growing number of intersex people, their families, activists, scholars, and some legal and health care professionals to question the medicalization of intersex. Here I use the term *medicalization* to refer to the treatment of human problems of the politics of difference as issues that can only be addressed by medical study, diagnosis, and treatment. As many scholars have argued, medicalization tends to reproduce, rather than resolve, a vast array of inequalities.[13]

According to intersex activists and scholars, except in rare cases where the infant's life is in jeopardy (such as the salt-wasting form of congenital adrenal hyperplasia, a potentially deadly metabolic condition), normalizing surgical and hormonal treatments are not medically necessary.[14] Just because a person is born with a body that looks different from what Audre Lorde calls the mythical norm does not mean that they require invasive medial treatment.[15] Regardless of intent, parental consent—or consent by proxy—does not resolve the bioethical conundrum of normalizing infant genital surgery.[16] To be sure, the majority of parents want what is best for their kids. They want their children to be able to live normal lives. Most physicians share the same goal. In a world that places such a high premium on normative genital appearance as a prerequisite for naturalized gender role ascription and acquisition, doctors ostensibly perform normalizing genital surgeries to humanize intersex infants—to allow them to be able to fit in. However, there is an unacknowledged cost to this way of thinking: surgical normalization reproduces the very stigma and trauma associated with intersex that physicians claim to prevent via surgery. Furthermore, there is no surgical or medical fix for what is essentially a social problem—the misapprehension of people born with atypical sexual anatomies as other, as defective, as disordered, as less than human. This misapprehension conditions the biomedical naturalization of the gender binary and the abiding cultural myth that humanity is "a perfectly dimorphic species."[17] In this regard, one might rightly conclude that intersexuality became an object of medical normalization not because it threatens the infant's life but because it threatens the infant's social milieu—specifically, the gender order status quo that is at once naturalized and reproduced by western biomedicine.[18]

The problem exposed by the medicalization of intersex is not, to borrow from Eve Kosofsky Sedgwick, that "people are different from each other."[19] Rather, the problem is "how few respectable conceptual resources we have for dealing this self-evident fact."[20] Since Sedgwick penned these axiomatic words more than twenty-five years ago, much has changed, yet intersex matters remain underaddressed in contemporary critical accounts of the politics of difference. In an effort to generate more respectable and capacious conceptual resources for dealing with the self-evident fact that bodies differ from each other, this book asks: How does intersex challenge and reconfigure contemporary understandings of sex, gender, and intersecting categories of difference and power such as race, class, sexuality, and nation? How might thinking critically about the norms, processes, and structures that regulate embodiment and subjectivity enable a critical rethinking of intersex, and vice versa? How do contestations over intersex shape, and how are they shaped by, the politics of difference and struggles for human rights and social justice in a multicultural, transnational world?

Addressing these questions, *Intersex Matters* proposes that intersex—as an analytic category, object of knowledge, and technology of subject formation—calls into question some of the principal dualisms (including male/female, sex/gender, mind/body, biology/culture, race/gender, nature/nurture, and normal/abnormal) that lend an illusory coherence to westocentric rationality and its globalizing structures of power. Analyzing the medicalization of intersex and the emergence of intersex rights activism in the global north and global south, I contend that intersex matters force us to rethink and further elaborate critical genealogies of biomedical embodiment, gender politics, and transnational activism; and I suggest that indispensible resources for this rethinking are to be found in the relatively decentered and decentering interdisciplinary perspectives of women's, gender, and sexuality studies (WGSS) and, especially, a queer feminist science studies approach informed by intersectional and transnational theoretical frameworks. *Intersex Matters* thus contests the view that intersex issues are solely relevant to a small sexual minority.[21] A critical attention to (how and why) intersex matters—the diverse ways that intersex lives, bodies, narratives, theories, and activisms materialize and become meaningful—profoundly reworks contemporary paradigms of embodiment, racialized gender and sexuality, health, normality, and human rights.

In the remainder of this introduction, I flesh out these claims by explaining how intersex poses particular challenges for dominant conceptualizations of embodiment, subjectivity, and identity politics. I then explain the interdisciplinary methodology that I develop for thinking through those challenges. Along the way, I provide some background on the history of intersex and also introduce some critical terminology to contextualize my argument as it develops across the book's chapters.

Intersex as the Problematization of Gender

Why do many people perceive intersex to be a problem in the first place? To answer this question, I turn here not to the history of biomedicine (as I do in chapter 1), feminist and queer theory (as I do in chapter 2), genealogies of intersex activism (as I do in chapters 3 and 4), or debates about sex testing in the world of international professional athletics (as I do in chapter 5), but rather to a first-person narrative. I begin by closely reading this narrative because it provides a vivid illustration of the rationale underlying the medicalization of intersex, as well as its material-semiotic assumptions, implications, and consequences.

In her 1998 autobiographical essay "Affronting Reason," Cheryl Chase recounts how, at the age of twenty-one, she was able with the help of a physician to obtain medical records from a hospitalization that occurred when she was an infant.[22] "It seems that your parents weren't sure for a time whether you were a girl or a boy," Chase's doctor explained. The doctor handed Chase her chart, which read " 'Diagnosis: true hermaphrodite. Operation: clitorectomy.' " "The hospital records showed Charlie admitted, age eighteen months. His typewritten name had been crudely crossed out and 'Cheryl' scribbled over it." Paradoxically, whoever crossed out "Charlie" and scribbled "Cheryl" over it (presumably one of Chase's physicians) had marked, in language, what the clitorectomy operation intended to hide or erase: that Chase was born intersex.[23]

The revelation of her hidden medical history was profoundly unsettling and traumatic for Chase. She writes,

> Though I recall clearly the scene of Dr. Christen handing me the records, dismissing me from her office, I can recall nothing of my emotional reaction. How is it possible that I could be a *hermaphrodite*? The hermaphrodite is a mythological creature. I am a woman, a lesbian, though I lack a clitoris and inner labia. What did my genitals look like before the surgery? Was I born with a penis?[24]

Chase explains that these questions literally brought her life into crisis.

> Fifteen years of emotional numbness passed before I was able to seek out the answers to these and many other questions. Then, four years ago, extreme emotional turmoil and suicidal despair arrived suddenly, threatening to crush me. *It's not possible*, I thought. *This cannot be anyone's story, much less mine. I don't want it. Yet it is mine.*[25]

Chase's exposition of the sheer *im*possibility of her life's story highlights the degree to which this story was imposed on her, from the outside, as it were, shaping her life chances in ways that were beyond her control. "*This cannot be anyone's story, much less mine*," Chase stresses. "*I don't want it. Yet it is mine*." These exclamations narrate a split. It is simultaneously a split in narrative and a split in subjectivity. Chase declares, on the one hand, that her story exceeds the terms by which we usually make sense out of human life, that her story seems to defy the basic conventions of narrative itself. On the other hand, Chase expresses that this impossible story, which she does not want, which she did not chose, is nevertheless her own. This co-constitutive split in narrative and subjectivity ultimately called the very possibility of Chase's existence and her capacity to go on living into question.

Confronting her medicalization as an infant head-on, Chase suggests that medical normalization has debilitating consequences for intersex people, their families, and the communities they inhabit. Moreover, she reveals that the biopolitics of sex and gender shape the ways doctors and parents treat intersex children in specific ways. For one thing, the clitorectomy rendered Chase anorgasmic, functionally eradicating her capacity for genital sensation and pleasure, which left a lasting mark on her adolescent and adult sexual development and relationships. In addition, medicalization transformed Chase's relationship to and understanding of her body and community. Chase writes,

> I learned that I had been born not with a penis but with intersexed genitals: a typical vagina and outer labia, female urethra, and a very large clitoris. Mind you, "large" and "small," as applied to intersexed genitals, are judgments that exist in the eye of the beholder. From my birth until the surgery, while I was Charlie, my parents and doctors considered my penis to be very small and with the urethra in the "wrong" position.
>
> My parents were so traumatized by the appearance of my genitals that they allowed no one to see them: no baby-sitters, no helpful grandmother or aunt. Then, at the very moment the intersex specialist physicians pronounced my "true sex" as female, my clitoris was suddenly monstrously large. All this occurred without any change in the actual size or appearance of the appendage between my legs.[26]

Chase's last observation in this passage is especially telling. Prior to her surgical sex reassignment, there first had to be a radical shift in the way her doctors and parents *perceived* Chase's infantile anatomy. Biomedical expertise transformed Chase's "small penis" into a "monstrously large clitoris," yet this transformation occurred "without any change in the actual size or appearance of" Chase's genitals. In short, being diagnosed as intersexual changed not only the meaning ascribed to Chase's body, but also its materiality. In stressing that " 'large' and 'small,' as applied to intersexed genitals, are judgments that exist in the eye of the beholder," Chase underscores that what counts as "normal" human genitalia is culturally specific and ideologically overdetermined. How we read the body is inextricable from normative gender schemas and our cultural assumptions about the nature of sexual and other forms of difference.

This point becomes even more pronounced when Chase describes how, after her surgical sex reassignment as female, as "Cheryl," her parents enforced a strict code of gender normativity. Just as psychoendocrinologist John Money (the founder of the modern medical approach to intersex, whose work I examine in chapter 1) recommended for other intersex infants, Chase's parents followed the doctors' recommendations and refused to openly and fully disclose to Chase her early medical history. "I know now," Chase writes,

> that after the clitorectomy my parents followed the physicians' advice, and discarded every scrap of evidence that Charlie had ever existed. They replaced all of the blue baby clothing with pink, discarded photos, birthday cards. When I look at grandparents, aunts, and uncles, I am aware that they must

> know how one day Charlie ceased to exist in my family, and Cheryl was there in his place.[27]

By underscoring that her extended family must be aware on some level that "one day Charlie ceased to exist in my family, and Cheryl was there in his place," Chase emphasizes her family's unacknowledged, silent, yet abiding complicity in the gendered regime of normality that conditioned Charlie's material-semiotic transformation into Cheryl.

For Chase's surgical sex reassignment from "Charlie" to "Cheryl" to have seemed medically urgent and necessary, her doctors and parents had to presume not only that the intersexual genitals Chase was born with were pathological, but also that surgery and resocialization could "correct" that pathology. This correction assumed that genital appearance is equivalent with anatomical sex, which is in turn equivalent with gender role and identity. This chain of equivalences was materially and semiotically inscribed onto Chase's body and personal history, and that inscription was profoundly traumatic for Chase. "Who am I?" Chase wonders in "Affronting Reason."[28] What precisely did the surgery do to Chase's body? And what did the surgery and its accompanying regime of normalization do to Chase's subjectivity and sense of self?

After fifteen years of personal struggle, it is perhaps no wonder that Chase found herself on the brink of suicide. Bravely stepping back from that brink, Chase turned instead to writing, organizing, and activism. She sought out and found other adults who had similar experiences as children. They formed bonds, born of shared trauma, and began to work together to transform medicine and social relations more broadly. In the process, Chase came to identify as both female and as "an avowed intersexual."[29] In 1993 she founded the world's first intersex activist organization, the Intersex Society of North America (ISNA), an organization that, as I show in this book, has profoundly impacted contemporary debates about intersex, biomedicine, gender politics, and human rights.

Crucially, Chase's activist career began with her critical recognition that it was ostensibly or so-called rational medical and social attitudes, and not intersexed peoples' bodies, that needed to change. In the conclusion of "Affronting Reason," Chase writes: "The time has come for intersexuals to denounce our treatment as abuse, to embrace and openly assert our identities as intersexuals, to intentionally affront the sort of reasoning that requires us to be mutilated and silenced."[30] In titling her essay "Affronting Reason," Chase figures intersex as a direct affront to reason, to common sense, to the status quo, and thus to the very modes of thought and practice that rationalize the medicalization of corporeal difference. Chase's autobiography deconstructs the westocentric rationalist calculus that seeks to hide and erase intersexuality with stigma, secrecy, shame, and nonconsensual genital surgeries. As Chase demonstrates, intersexuality is an affront to reason, and affronting reason is a transformative epistemological, personal, political, and ethical act.

"The fact that my gender has been problematized," Chase writes in a sentence whose importance for the present inquiry cannot be understated, "is the source of my intersexual identity."[31] During the last half-century, western biomedical practitioners have treated intersexuality as the de facto *cause* of a gender problem. From the dominant medical perspective, the diagnosis of intersexuality means that the infant's sex is "unclear" and that their gender may be or become "ambiguous." Surgical normalization proceeds on the basis of this assumption. Reversing that conventional logic, Chase figures her intersexual identity not as a purely inborn state or condition of pathology, but rather as the result of a complicated material-semiotic process: the problematization of her gender. When she was born, Chase implies, her gender was not in and of itself a problem. After all, at birth infants' genders, which are distinct from and not necessarily determined by their sexed morphologies, are just beginning to form.[32] It is impossible to predict how any infant's gender identity will congeal and/or change over time. Yet in the case of infants like Chase, doctors and parents promptly *decide* that the bodies they were born with are inadmissible and unintelligible and then immediately *turn* their gender into a problem.

This point is absolutely decisive. Intersexuality, Chase reveals, is not problematic in and of itself. Rather, what is problematic is a biomedical regime of knowledge and power that is incapable of interpellating intersexuality as anything but a problem. In other words, Chase shows that intersexuality only becomes a figure or occasion for what Butler calls gender trouble—the trouble that emerges when gender and the power relations that underlie its presumed meaning and materiality are called into question—in a social system that enforces a strict code of binary gender normativity and sexual dimorphism.[33]

The Biopolitics and Geopolitics of Intersex

Chase's narrative sets the stage for this book's central thesis: that intersex challenges us to rethink and rearticulate the biopolitics and geopolitics of sex, gender, race, and nation in new ways. First theorized by Michel Foucault, biopolitics refers to the intersection of modes of living with techniques for managing bodies and populations.[34] Biopolitics takes shape through myriad social, political, economic, and cultural institutions, such as health care and legal systems, government, media, kinship, nongovernmental organizations, capitalism, and international agencies, as well as through more informal discourses, networks, and practices. I use the term geopolitics in a related sense to refer to the situated national and transnational power dynamics of various types of local and global flows. As Inderpal Grewal argues, biopolitics and geopolitics not only intersect and mutually inform one another in the era of neoliberalism, but also are profound sites of struggle and contestation.[35]

ISNA, the activist organization that Chase founded, recognized as much when it defined intersex as "people born with an anatomy that someone decided is not standard for male or female."[36] ISNA's definition of intersex suggests that the categories *male* and *female* are not self-evident objective labels whose meaning is transparent and identical across all times and places. ISNA asserts that the criterion for determining which bodies count as *male* or *female* is not anatomy alone, but rather what "someone decided" gets to count as "standard" for anatomy. A growing body of interdisciplinary scholarship corroborates ISNA's hypothesis. Definitions of male and female anatomy are not predetermined or universal but are instead culturally specific and contextually standardized.[37] That is, they are historically contingent, relational, shifting, and to various degrees arbitrary. This also helps to explain why genealogies of racial difference are absolutely central to normative definitions of sex dating all the way back to the onset of European colonization and extending into the present (more about this point below).[38] Though many people are accustomed to thinking of the male/female distinction as the most scientific and natural division of humankind, ISNA suggests something quite different and quite illuminating: that labeling someone as male, female, cisgender, trans, or intersex is a social *decision*, not an impartial, neutral description of an empirically verifiable biological state. Crucially, the "someone" ISNA names as responsible for this decision is anonymous: it could be *anyone*. ISNA thereby implicates a broad range of parties, not only doctors and parents, as accountable for participating in the stigmatization and exclusion of people born with anatomies deemed to be nonstandard from belonging to the domain of the intelligibly human. This stigmatization and exclusion calls for critical reflection on the place of intersex in the contemporary biopolitical and geopolitical order of things.

Intersex and the Sex/Gender Distinction

In addition to figuring sex/gender categories as socially, semantically, and materially consequential decisions, ISNA's definition of intersex puts critical pressure on a distinction that became, during the twentieth century, widely accepted in western biomedical discourse and also in the English-speaking world more broadly: the sex/gender distinction. This distinction has informed mainstream understandings of gender as masculine or feminine identification and expression and has also been used as a grounding assumption of the late twentieth-century feminist articulation of the social construction of gender. According to the conventional understanding of this distinction, sex is biological. It is said to consist of the chromosomal, gonadal, hormonal, internal morphogenic, genital, and possibly neurocognitive features that characterize the dimorphic division of the human species into males and females. Gender, by contrast, is defined as social and cultural. It is said to refer to the *learned* behaviors, roles, identifications, and ways of being and acting in the world that societies differentiate and distribute asymmetrically along the masculine/feminine axis. In this paradigm, male and female are sex categories, whereas man and woman, and boy and girl, are gender categories. Many people assume that there is a natural alignment and causality between sex and gender, such that one's gender is assumed to essentially follow from one's sex. This line of thought holds that biological *males* are supposed to develop *masculine* gender identities, and biological *females*, *feminine* gender identities. From this standpoint, when a child is born with an anatomy that does not readily conform to normative expectations about sexed embodiment, the child's prospective identity and status as a coherently gendered subject is thrown into question. Based on an epistemological paradigm that opposes nature to culture, the dominant reiteration of the sex/gender distinction accords to sex a status so foundational as to make subjectivity itself seem unimaginable in absence of the dimorphic schema of sex.

Intersex bodies are trouble, then, not because they are inherently "disordered," but because they reveal that sex categories—in this instance, male, female, and intersex—are overdetermined, incoherent, and unstable. From this perspective, what counts as intersexuality materializes in relation to and against normative conceptions of male and female embodiment. As Karkazis, drawing on Morgan Holmes, argues, "whatever intersexuality may be physiologically (and it is many things), intersexuality as a category of person (requiring medical treatment) is not natural."[39] As such, intersex is, like male and female, a situated naturecultural interpretation of particular bodily forms.[40] Yet, unlike male and female, intersex is presumed (by dominant institutions) to be not only nonstandard, but also abnormal. For particular biopolitical and geopolitical reasons that this book investigates, intersexuality has become one of the key "others" against which hegemonic notions of dimorphic sex and binary gender have been defined.

Queer Feminist Science Studies, Intersectionality, and Transnational Feminisms

At its core, *Intersex Matters* suggests that the medical regulation of atypical sex is fundamentally a feminist and queer issue, and an intersectional and transnational one as well. Throughout the book's chapters, I track how a specific category of human embodiment—intersex—underwent key material-semiotic shifts over the course of the twentieth and early twenty-first centuries. During this time period, intersex went from signifying an anatomical abnormality that could be medically corrected in the 1950s through the 1980s, to a reclaimed, oppositional, and politicized queer dis/identification in the 1990s, to a transnational human rights issue in the 2000s, to a supposedly outdated label that now, in the second decade of the twenty-first century, is being replaced by the medical term *disorders of sex development*. Of course, these shifts were far messier than this brief outline makes them seem. In many instances, these models continued to exist alongside and in tension with one another, rather than being subsequent, linear, developmental transitions. Nonetheless, each of these shifts has been marked by intense contestation. To analyze these contestations, *Intersex Matters* crafts a methodology of reading intersecting biopolitical and geopolitical processes to expose changing configurations of power and knowledge. The methodology I develop is what I call a queer feminist science studies approach informed by intersectional and transnational perspectives. This approach provides especially useful tools for rethinking the assumptions underlying the scientific codification and biomedical regulation of bodies and also helps us to understand the challenges to such regulation occasioned by activist, political, and scholarly critical practices.

The argument *Intersex Matters* develops is in conversation with recent feminist and queer scholarship on new materialism, posthumanism, and the nonhuman turn.[41] I am indebted to and have learned much from these innovative bodies of research. At the same time, I share Sara Ahmed's concern that some of the founding rhetorical gestures of new materialism and posthumanism may erase earlier feminist and queer work on embodiment, matter, and biology that "emphasizes precisely the entanglements and traffic between nature/biology/culture and between materiality and signification."[42] Rather than positioning queer feminist science studies against new materialism and posthumanism, I would suggest that queer feminist science studies provides a rich, underexplored alternative intellectual genealogy for thinking dynamically, developmentally, and critically about the relationships between embodiment, materiality, and ecosystems in ways that take intra-acting biological and cultural factors into account.[43] It is my hope that this book might productively contribute to current and future interdisciplinary dialogues about materiality and biocultures by showcasing the interpretive resources and analytic possibilities opened up by queer feminist science studies. In this sense, *Intersex Matters* seeks to think materiality in new ways.

In addition, my methodology is informed by the intersectional and transnational perspectives that have become central to interdisciplinary research in the field of WGSS and beyond.[44] Coined by Kimberle Crenshaw and theorized in US women of color scholarship, particularly black feminist thought, intersectionality refers to the co-constitution of systems of oppression based on race, class, gender, and other categories of difference.[45] Intersectional theorizing apprehends systems of oppression as interlocking, as the Combahee River Collective famously put it.[46] Within the genealogies of US women of color and queer of color critique—the latter being heavily influenced by the former—intersectional approaches foreground the inextricability of race from gender, class, and sexuality. At the same time, they also analyze how intersectional subjects, such as "women of color," are uniquely disenfranchised by the unequal distribution of life chances in white supremacist, capitalist, heteronormative, and patriarchal societies.[47]

Relatedly, transnational feminist perspectives reveal that struggles over citizenship, cross-border politics, globalization, and the international division of labor are key concerns for any scholarship and activism engaged with the production and reproduction of bodies.[48] Transnational feminist approaches focalize local-global interdependencies and explore the diverse ways in which dissimilar and uneven circuits of capital and culture shape gendered, racialized, and classed subjects.[49] As Karen J. Leong, Roberta Chevrette, Ann Hibner Koblitz, Karen Kuo, and Heather Switzer argue, "bringing together intersectionality and transnational feminisms . . . sharpens the analytic potential of each."[50] Conjoining queer feminist science studies with intersectional and transnational perspectives, I contend that intersex bodies and histories materialize through diverse genealogies of biopolitics, gender regulation, racialization, citizenship, and geopolitics.

Complicating Intersex

Intersex Matters draws on these methodological and interpretive frameworks to propose that intersex embodiment is shaped by even as it troubles dominant discourses of sex and gender as they intersect with heterogeneous formations of sexuality, race, nation, class, and ability. My approach begins by questioning the notion of sex itself. According to Foucault, the notion that each person possesses a "true sex" has been pivotal to the disciplinary production of the modern subject.[51] In a frequently cited passage from the English translation of the first volume of *The History of Sexuality*, Foucault argues that

> the notion of "sex" made it possible to group together, in an artificial unity, anatomical elements, biological functions, conducts, sensations, and pleasures, and it enabled one to make use of this fictitious unity as a causal principle, an omnipresent meaning: sex was thus able to function as a unique signifier and as a universal signified.[52]

Reading the above passage, Butler suggests in *Gender Trouble* that,

> for Foucault, the body is not "sexed" in any significant sense prior to its determination within a discourse through which it becomes invested with an "idea" of natural or essential sex. The body gains meaning within discourse only in the context of power relations. Sexuality is an historically specific organization of power, discourse, bodies, and affectivity. As such, sexuality is understood by Foucault to produce "sex" as an artificial concept which effectively extends and disguises the power relations responsible for its genesis.[53]

Neither Foucault nor Butler implies that sex is merely a product of language or discourse. Rather, both suggest that the category sex only becomes *meaningful* in and through a discursive framework that effaces its own historicity and artificiality. This does not mean that sex is immaterial; it simply means that sex materializes, in part, in and through cultural and discursive practices. Butler's interpretation frames "sex" as an ideological backformation or retroactive effect of what Foucault calls the deployment of sexuality, which Butler retheorizes in terms of the cultural logic of binary gender. Butler thus uses Foucault to argue that cultural gender precedes, produces, and undoes the "fictitious unity" of biological sex.[54]

Butler's work has been tremendously influential in bringing "French poststructuralism" to bear on "American feminist theory," but critics have perhaps paid less attention to ways in which *Gender Trouble* participates in what we might call the Americanization of Foucault. In her 2010 monograph *Mad for Foucault: Rethinking the Foundations of Queer Theory*, Lynne Huffer suggests that it is crucial to attend to the linguistic specificity of

Foucault's argument in the first volume of *The History of Sexuality*.[47] In the original French version of the above passage from *Volume One*, Foucault uses the term *le sexe*. Huffer contends that the rendering of Foucault's *le sexe* into the English, and specifically American, rubric of "sex" raises "a problem of translation."[55]

> As its linguistic ambiguity in French suggests, the "dense transfer point of power" Foucault calls *le sexe* includes within it all the meanings English speakers differentiate into sex-as-organs, sex-as-biological-reproduction, sex-as-individual-gender-roles, sex-as-gendered-group-affiliation, sex-as-erotic-acts, and sex-as-lust. And if *le sexe* is produced by the dispositif of sexuality, this hardly means it supersedes or reverses the primacy of gender, as many queer theorists would like to claim. Sex, sexuality, and gender are inseparable and coextensive.[56]

Challenging queer accounts that privilege sexuality above gender, as well as feminist accounts that privilege gender above sexuality, Huffer usefully troubles the queer/feminist split in contemporary scholarship, reading sex, sexuality, and gender as mutually constitutive categories.

Bringing Huffer's intervention into conversation with the history of science, specifically critiques of sexology and eugenics (the fields that directly shaped the modern medical paradigm of intersex treatment), *Intersex Matters* hones in on the specific historical relationships between intersexuality, the sex/gender distinction, and processes of sexual, racial, class, and national formation in particular contexts. Following Sally Markowitz, I argue that unmarked references to sexual dimorphism, gender, sexuality, and indeed intersex too can reify the presumed whiteness of those categories.[57] Theorizing as intersexuality an effect of "institutions, practices, [and] discourses with multiple and diffuse points of origin,"[58] *Intersex Matters* puts pressure on the presumption that intersex is only an issue of stigma and trauma (as ISNA argued), or solely of gender (as US feminist and queer scholars have argued), in order to read intersex intersectionally and transnationally. Without a doubt, intersex *is* an issue of stigma and trauma *and* of gender, but it is *also* fundamentally an issue inseparable from the politics of difference more broadly—especially differences of race, class, sexuality, ability, and nation. Through this intersectional, transnational, and diffractive reading practice, I argue that the medical management of intersex can be conceptualized as central to the working out and ongoing regulation of hegemonic and subordinate configurations of the human in the twentieth and twenty-first centuries.

Chapter Overview

Before outlining the chapters that follow, a brief explanation of the relationship between the transnational feminist dimension of my methodology and the structure of the book is in order. Some readers' comments on earlier versions of this manuscript seemed to assume that a transnational approach is equivalent to an international or global one—that they all basically refer to studies of things located beyond the geographic borders of western nation-states. This is a common misunderstanding of transnationalism, especially in an era wherein the US academy has realized the strategic political economic value of "globalizing" its curriculum so as to produce "global citizens."[59] Such an approach tends to take the west/non-west, domestic/foreign, developed/developing, first world/third world, and global north/global south distinctions for granted as objective descriptions of supposedly transparent empirical realities. By contrast, the transnational feminist approach I employ challenges the naturalization of these binaries. It does so by foregrounding the historical interdependencies and uneven and asymmetrical power relations between nation-states. The local is never simply local but always infused with global influences, and the global always materializes in and through localized political and economic relationships.[60] The US nation-state is therefore transnational in this sense, in that its material-semiotic borders have been irrevocably shaped by histories of imperialism; settler colonialism; racial capitalism; forced, coerced, and free migration; environmental depletion; and other planetary life-altering processes.

I say all this to explain why, even though the chapters of the book move from US to Colombian and South African examples, this structure is expressly not intended to suggest a teleological line of development from the "west" outward to the international or global scene. On the contrary, the book seeks to contribute in whatever small way it can to the undoing of the world historical processes that are its conditions of possibility, including US exceptionalism and its imperial legacies. The first three chapters focus on examples from North America because, plainly put, that's where the medicalization of intersex and its activist and academic critiques first emerged. The final two chapters focus on Colombian and South African case studies respectively in an effort to register global southern resistances to the rapidly amassing global dominance of westocentric biomedical and activist approaches to atypically sexed bodies. My aim, then, is not to offer a comprehensive analysis of intersex treatment, activism, and theory in every nation around the globe (a project that far exceeds the parameters of any single monograph), but rather to use these particular examples to rethink the biopolitics and geopolitics of intersex in specific contexts that have far-reaching consequences.

Chapter 1 sets up the genealogical and theoretical framework of the book by tracing an account of intersexuality's underrecognized but historically pivotal role in the development of gender as a concept in twentieth-century western biomedicine, feminism, and their globalizing circuits. I argue that intersex has been and remains central to the history of gender as a classificatory schema, an object of knowledge, a technology of subject formation, and a paradigm of sociality in late modernity. This genealogy pushes beyond current scholarship on intersexuality to suggest that, while dominant understandings of sex and gender have overdetermined the meaning of intersex, historically speaking, the concept of intersex paradoxically preceded and inaugurated what we would today call the sex/gender distinction. Through a close reading of research by psychoendocrinologist John Money (the principal architect of the dominant medical paradigm of intersex management), I show that intersex was integral to the historical emergence of the category *gender* as distinct from *sex* in the mid-twentieth-century English-speaking world. I argue that Money used the concept of gender to cover over and displace the biological instability of the body he discovered through his research on intersex, and that his conception of gender produced new technologies of psychosomatic normalization. I critique the heteronormative masculinism and unmarked whiteness of Money's approach to intersex, but I also examine his lasting yet underinterrogated legacy in feminist scholarship from the second wave through the present.

Chapter 2 analyzes the past two decades of feminist and queer scholarship on intersexuality to suggest that the study of intersex does not merely represent an expansion of the range of objects studied in gender studies, but in fact usefully troubles and creates opportunities to rethink interdisciplinary analytics of sex/gender, sexuality, embodiment, and the politics of difference more broadly. In so doing, I contend that the critical tools of feminist and queer thought make it possible to apprehend and preserve what Iain Morland calls the uncertainties intersex bodies provoke. Closely reading foundational works by Kessler, Fausto-Sterling, and Butler, I argue that their analyses of intersex productively rethink the sex/gender, nature/culture, and gender/sexuality distinctions, interrupting heteronormativity's equation of genitalia with sex and gender. In addition, this chapter addresses the problems of intersex exceptionalism and the fetishization of intersex alterity as they play out in Kessler, Fausto-Sterling, and Butler and reflects on the role these problems continue to play in contemporary critical intersex studies.

Chapter 3 traces a genealogy of the intersex movement in the United States and reconsiders intersex critiques of women's studies and gender studies, queer theory, and feminist scholarship. I examine these critiques within writings by well-known US-based intersex activists and scholars, including Cheryl Chase, Alice Dreger and April Herndon, Emi Koyama and Lisa Weasel, Vernon Rosario, and Katrina Karkazis. These writers critique feminist and queer scholarship for appropriating intersex for gender theory and for overinvesting in theoretical gender at the expense of the lived experiences of intersex people. Closely reading the ways in which these writers routinely reiterate, but do not interrogate, a central claim made by ISNA—that "intersex is primarily a problem of stigma and trauma, not gender"—I argue that this claim not only ignores the genealogical links (which I trace in chapter 1) between the mid-twentieth-century medicalization of intersexuality and the emergence of the modern bioscientific sex/gender distinction, but also paradoxically figures the materiality of intersex experience as somehow completely divorced from gender regulation and other manifestations of biopower. Contra these writers, I suggest that feminist and queer thought provide capacious interpretive resources for situating the medicalization of intersexuality and activist responses to medicalization in relation to biopolitics and geopolitics. Most crucially, feminist and queer theory enable us to recognize the limitations of neoliberal models of subjectivity and political reform for imagining intersex otherwise.

Chapter 4 extends and deepens the genealogy of intersex activism addressed in chapter 3. It explores how debates about intersex are shaped, challenged, and interrupted by geopolitical power relationships, global activism, and transnational feminist perspectives. It does so by examining two crucial events in the history of intersex activism: ISNA's failed attempt to lobby for the inclusion of unnecessary intersex surgeries in the US Congress's 1997 federal ban on "female genital mutilation" (FGM); and ISNA's influence on two 1999 decisions by the Constitutional Court of Colombia to rework the definition of informed consent and to limit doctors' capacity to perform normalizing genital surgeries. In my analysis of these events, I pay special attention to how human rights discourse, US imperialism, biopolitics, and neoliberalism contour the locational politics of intersex activist projects situated in different national contexts. Concurrently, I suggest that the transnational regulation of sexed bodies occurs not only through the globalization of western biomedical models of sex/gender normativity, but also through the circulations of human rights discourse—and through the imposition of US neoliberal democratic frames of subjectivity. My reading of the Colombian case in particular focuses on how these frames travel across national borders and solidify (even as they are challenged and interrupted by local actors) and reveals the consequences of this movement. This chapter therefore concludes by analyzing the ways in which intersex activists and scholars in Colombia and other nations in the global south have challenged the US-centrism, unmarked whiteness, and imperialism of western/global northern intersex human rights campaigns.

Chapter 5 further excavates the relationship between intersectionality and intersex in transnational times. To a large degree, the field of intersex studies has focused on theorizing intersex in terms of the ethics and politics of surgical normalization. While this focus has revealed key insights into what ISNA called the "stigma and trauma" caused by the medicalization of intersex bodies, it does not sufficiently account for the role of intersecting formations of racialized gender, nation, and sexuality in the medical and social management of intersex in different geopolitical contexts. Likewise, intersectional scholarship has not addressed whether and how intersex might reconfigure the grounding presuppositions of intersectionality as a theoretical framework. Addressing these lacunas, this chapter asks how intersectional perspectives might help us to expand, complexify, and refine understandings of intersex, and how intersex might likewise enable a rethinking of the premises of intersectionality.

To do so, this chapter pursues a discursive analysis of media representations and scholarly and activist accounts of South African professional middle-distance runner Caster Semenya. Semenya became a household name around the globe in 2009 after she was required by the International Association of Athletics Federations and the International Olympic Committee to undergo "gender testing" by an interdisciplinary medical team. Despite the fact that Semenya is a woman and has never publicly identified as intersex, my analysis suggests that contestations over Semenya's story are as much about racialized gender and sexuality, imperial history, and national context as they are about intersex and the politics of sex and gender regulation. For this reason, I argue that Semenya's story was never entirely *her* own story at all. Rather, Semenya's story became a history of the appropriation of her story for the agendas of western and non-western elites. This chapter therefore concludes by considering how Semenya herself spoke back to these appropriations and thereby implicated athletic officials and fellow competitors, fans, politicians, scholars, and activists in the silencing of subaltern speech.

In the conclusion, I reflect on a 2005 proposal by a widely recognized coalition of western-trained medical experts and activists called the DSD Consortium, which proposed to rename intersex conditions with the acronym DSD (disorders of sex development). The medical profession has rapidly embraced the DSD nomenclature, but some activists and scholars have raised concerns over its political and ethical implications. Situating the DSD debates in the context of contemporary reterritorializations of empire, processes of neoliberal restructuring, and the retooling of biopolitical technologies of corporeal regulation, I argue that the DSD nomenclature can be understood as a medico-scientific attempt to pin meanings and bodies down and to control and obscure the uncertainties about embodiment that intersex bodies expose. By way of conclusion, I propose an ethics of uncertainty as a vital resource for articulating alternatives to the medicalization of people of atypical sex and, ultimately, for thinking intersex/DSD otherwise.

Although I am a rather private person and this is not a work of autobiographical disclosure, I feel that it is pertinent to note here that a number of years ago I was diagnosed with a DSD. I am not going to tell you the details of my specific diagnosis, as I reserve my right to keep my medical history private. What I will tell you is that, after some additional tests, my physician recommended a normalizing course of treatment. When I asked whether the treatment was necessary for the sake of my health and wellness, my doctor equivocated. When I pressed him, he eventually relented, telling me that, although there was no impending threat to my physiological health, the treatment would make me look and feel more like a normal guy. What, I wondered then and wonder now, does a "normal guy" look and feel like? Is there really only one template?

What I found objectionable in my doctor's medical recommendation was the assumption that fitting in is better than not knowing precisely where one fits in the first place. Because I was an adult at the time of my diagnosis, I was given a choice as to my treatment options. But the "choice" I was given was, at least in my doctor's eyes, really no choice at all. My doctor didn't give a second thought as to whether medicine should be in the business of adjudicating gender normativity, or whether normalization itself is the preferable psychosocial good. My decision to decline treatment positioned me as a liminal subject, a bad patient, and a gender outlaw. In another, slightly earlier historical moment, I might have been medically interpellated as a person with an intersex condition. And in an even earlier period, my physical difference might not have been noticed at all. This experience raised numerous questions for me about the material-semiotic status of my body, subjectivity, identity, and work, questions that I am still asking today. Ultimately, it reinforced for me the importance of questioning the gendered, classed, racialized, and nationalized logics underlying the biomedical regulation of the distinction between people with and people without DSD, a distinction whose intelligibility depends on the very medicalization of intersex that it also effaces/replaces. As someone who seeks—in my scholarship and teaching—to be an ally to the intersex movement, this experience gave me a firsthand understanding of what it is like to be medically othered and pathologized in service of the maintenance of the mythical norm.

Although *Intersex Matters* advocates for the ethical value of bodily diversity, my point isn't merely that we need more categorical options, but rather that any manner of categorizing bodies will be inextricable from historically and bioculturally situated configurations of the politics of difference. This is why self-reflexivity and critical positionality matter to the politics of the production of knowledge. In the current conjuncture, efforts to denaturalize dualism and dimorphism must also recognize that these are not the only manifestations of power. Normalizing regimes take many forms, but none is ever finally totalizing. As the genealogies that I trace herein suggest, the challenge is to keep normalizing processes open to critical contestation in order to animate collective curiosity and democratic debate about how the world might be worlded otherwise.[61]

1

"An Unnamed Blank That Craved a Name"

A Genealogy of Intersex as Gender

This chapter traces a genealogy of intersexuality's underrecognized but historically pivotal role in the development of gender as a concept in twentieth-century American biomedicine, feminism, and their globalizing circuits. According to Michel Foucault, genealogy "rejects the metahistorical deployment of ideal significations and indefinite teleologies."[1] Genealogy opposes itself to the search for monocausal origins. As a critical methodology, it focuses instead on the conditions of emergence and force relations that shape diverse and discontinuous embodied histories. The task of genealogy, Foucault writes, is to "expose a body totally imprinted by history and the process of history's destruction of the body."[2] As an analysis of the will to knowledge, genealogy reveals the exclusions by which dominant historical formations constitute themselves and focalizes the roles of interpretation and "the hazardous play of dominations" in the materialization of bodies in particular spaces and times.[3] Genealogy, then, proves immanently valuable for a queer feminist science studies project informed by intersectional and transnational perspectives.

Contrary to the view that intersex is only relevant to a small sexual minority, this chapter suggests that the western medicalization of intersex centrally shaped the very idea of gender as a generalizable rubric for describing what came to be seen, starting in the mid-twentieth century, as a core, fundamental aspect of human intelligibility: self-identification and expression as masculine or feminine. The category of gender found quick uptake in both the production and contestation of other intersectional hierarchies of difference, especially those of race, class, sexuality, ability, and nation. In the global north and its scattered hegemonies,[4] western biomedical understandings of natural sexual dimorphism and normative gender were grounded in unmarked ideological investments in whiteness, heteronormativity, bourgeois-status, and compulsory able-bodiedness.[5] The genealogy I offer calls into question the idea that gender provides the most primordial, accurate, or extensive language for accounting for human differences. On the contrary, gender is a historically and geopolitically situated technology of subjectivation and subjection whose intelligibility as a binary system is materially contingent not only on the erasure of the racializations of sexual dimorphism in western science and culture, but also on yet *another* erasure: of the medicalization of intersex, trans, and gender-nonconforming subjects.

In myriad underanalyzed but absolutely world-shaping ways, the genealogy of intersex conditions understandings of gender as a classificatory schema, object of knowledge, technology of subject formation, and paradigm of sociality in late modernity. While dominant mappings of sex and gender have overdetermined the meaning of intersex, historically speaking the concept of intersex paradoxically preceded and inaugurated what we would today call the sex/gender distinction. Through a close reading of psychoendocrinologist John Money's biomedical research, I show that intersex was integral to the historical emergence of the category *gender* as distinct from *sex* in the mid-twentieth-century English-speaking world. In this chapter, I critique the heteronormative masculinism, westocentrism, and raced and classed presuppositions of Money's approach to intersex, but I also examine his lasting yet underinterrogated legacy in feminist scholarship from the second wave through the present.

Gender in Intersex Studies, Feminist Theory, and Biomedicine

In 1990, Suzanne J. Kessler published "The Medical Construction of Gender: Case Management of Intersexed Infants" in *Signs: Journal of Women in Culture and Society*.[6] Ranked among the top-twenty most cited *Signs* articles of the past two decades, Kessler's essay focalized a practice that was, until the early 1990s, rarely discussed outside specialized medical circles: the surgical normalization of infants born with sexual anatomies deemed to be nonstandard. Analyzing interviews with physicians and the medical literature on intersex treatment, Kessler argued that "members of medical teams have standard practices for managing intersexuality that ultimately rely on cultural understandings of gender."[7] In making this claim, Kessler pointed to the significance of clinicians' reliance on what is known as the optimal gender paradigm. Developed by Money and his various colleagues over the years, I examine this paradigm in greater detail below. Here, it suffices to note that the optimal gender paradigm is a treatment model that seeks to help physicians select the most optimal gender for individuals born with atypically sexed anatomies. Its central presumption is that surgical normalization can and should be used to foster the development of conventional gender identities. Kessler, however, was concerned about the ethical implications of this paradigm, specifically the ways in which it medicalized intersexual difference so as to maintain the gender order status quo. Noting that the vast majority of intersex "conditions" pose little or no health risk, she concluded that intersexuality "is 'corrected' not because it is threatening to the infant's life but because it is threatening to the infant's culture,"[8] an argument she would reiterate in her 1998 monograph *Lessons of the Intersexed*.[9]

Following Kessler's lead, during the past twenty-five years a small but growing number of scholars have made vital contributions to feminist and queer theory, science studies, bioethics, medical sociology, and debates about human rights and bodily integrity by showing that intersexuality challenges naturalized understandings of embodiment through analyses of the medical construction of sexual dimorphism.[10] As my language indicates, I am interested in the implications of an unremarked discursive shift that began to manifest itself as this body of interdisciplinary research developed. In Kessler's wake, the analytic preoccupation of intersex studies was displaced almost immediately from gender to sex, as evidenced by the titles of works published following "The Medical Construction of Gender" such as *Sexing the Body*, "Sexing the Intersexed," and *Fixing Sex*.[11] One could interpret this shift as a transition from the social back to the somatic, reading the emphasis on *sexing* as consonant with and influenced by the recognition of the limitations of the essentialism/constructionism divide and the consequent push to rethink the materiality of the body in 1990s feminist theory post–*Gender Trouble*.[12] But this alone does not explain why gender receded into the theoretical background of intersex studies as the field began to congeal. While Anne Fausto-Sterling, Sharon Preves, Katrina Karkazis, and others have undoubtedly sharpened critical perspectives on the medical and social treatment of people with intersex embodiments, and while their work hints at the import of gender as a system of power, their accounts have largely focused on rethinking the sex side of the sex/gender distinction. For this reason, less attention has been paid to questions about the genealogical relation between intersex and gender, questions that were implicitly posed but not fully answered in Kessler's initial *Signs* essay, such as: What is the historical relationship between intersex and the sex/gender distinction? How has the sex/gender distinction shaped and been shaped by intersex?

The elision of these questions has been reinforced by an influential strain of intersex activism. Throughout the 1990s and 2000s, the Intersex

Society of North America (ISNA) avowed that "intersexuality is primarily a problem of stigma and trauma, not gender."[13] With this claim, ISNA sought to reframe the terms of medical, scholarly, and popular discourses about intersex. As Iain Morland argues, this claim crucially "acknowledged that affected individuals—rather than their parents or doctors—are experts on their own genders," and further suggested "that traditional treatment . . . often inadvertently creates trauma and thus fails by its own standards."[14] In this way, ISNA challenged the medical model of intersex management, and that model's surgical equation of dimorphic genitalia with normative sex, promoting instead a patient-centered approach founded on intersex adults' critical reflections on their experiences of medicalization. Although this claim buttressed ISNA's opposition to nonconsensual genital surgery, it also obscured and, because of ISNA's lasting impact, continues to obscure the powerful role of gender in the development of modern intersex medicine and the sciences of sexual health more broadly. Before it became a key term in feminist discourse, and before it came to signify the social construction of femininity and masculinity, gender was formulated in mid-twentieth-century American sexology as a diagnostic solution to the so-called medical emergency of intersex bodies, or bodies in doubt, to borrow historian Elizabeth Reis's apt characterization.[15] The story of intersex is therefore not only, as ISNA asserted, a story about "shame, secrecy, and unwanted genital surgeries," a story about "stigma and trauma," but also a story about the regulation of embodied difference through biopolitical discourses, practices, and technologies of normalization that materialize in, through, and as gender.[16]

Contemporary intersex activists and scholars have taken Money to task for his perpetuation of intersexism and heteronormativity.[17] Paradoxically, however, as the medical model of intersex management that his research institutionalized came under fire in the 1990s and 2000s, the role of gender in that model has been dissimulated. While the medical and social treatment of people with intersex is not reducible to gender dynamics alone (as I argue in the latter chapters of this book), the significance of gender for intersex, and vice versa, has yet to be fully recognized.

As Jennifer Germon argues, gender does in fact have a history, and "a controversial one at that."[18] In *Gender: A Genealogy of an Idea*, Germon draws on Bernice Hausman to argue that it was not until the mid-twentieth century that English speakers began using gender as an ontological category, a category said to denote masculine and feminine states of subjective being.[19] In particular, Germon suggests that Money's influence on the career of the gender concept has been even more decisive than Kessler initially indicated. According to Germon, it was through the research Money undertook as a graduate student at Harvard University in the 1950s on hermaphroditism (a term he used interchangeably with intersex), and subsequently pursued treating intersex patients at Johns Hopkins University, that the gender concept came to be recognized as an explanatory measure of human behavior in the biomedical and social sciences.

Germon's analysis of Money is not only critical but also reparative. To the degree that Money has become the proverbial archnemesis of the intersex movement, and insofar as scholars in intersex studies have sought to support intersex activists' arguments against coercive genital surgery, Money has been frequently criticized but less often read closely. Germon counters this tendency, suggesting that Money was scarcely the hard-line constructionist his detractors paint him to be. In addition, Germon argues that Money's ideas, despite their problematic investments in medical paternalism and the binary model of sexual difference, nevertheless manifest a strong interest in understanding nature and culture within a more complex interactionist framework.

In their 2015 book *Fuckology: Critical Essays on John Money's Diagnostic Concepts*, Lisa Downing, Iain Morland, and Nikki Sullivan challenge Germon's optimism that Money's work adopts an interactionist approach to the relationships between the cellular, environmental, and experiential domains.[20] Through both cowritten and individually authored close readings of various parts of Money's oeuvre, Downing, Morland, and Sullivan show that Money's claims were often conflicting, self-undermining, and dysfunctional. For instance, Sullivan contends that Money was neither an essentialist nor a constructionist in any simple or straightforward sense, but does suggest that Money's model of gender "posits the biological as foundational" to the subsequent development of psychosexual gender identities and roles.[21] Doing critical justice to Money's vast and contradictory body of research and its legacies, she concludes, requires that we "trouble the tendency to see in dimorphic terms."[22]

My analysis converges with Downing, Morland, and Sullivan's and Germon's in exploring the enduring significance of Money's research, and the centrality of the intersexed to the history and politics of gender. In an effort to deepen and extend these analyses, I argue not only that intersexuality played a crucial role in the invention of gender as a category in mid-twentieth century biomedical and, subsequently, feminist discourses; and that Money used the concept of gender to cover over and displace the biological instability of the body he discovered through his research on intersex; but also that Money's conception of gender produced new technologies of psychosomatic normalization. In contrast with Germon, my aim is not "to critically reinvigorate Money's gender"[23] concept but rather, following Downing, Morland, and Sullivan, to more fully excavate the broad swathe of its regulatory power.

Rethinking Sex and Gender

With the exception of Germon, Hausman, Downing, Morland, and Sullivan, and a few other scholars,[24] the impact of Money's research on the development of the modern conception of gender has not received sustained feminist attention. This is striking because intersex has been linked to gender in feminist discourse since at least the early 1970s. In her 1972 monograph *Sex, Gender and Society* (which has been out of print for many years), British sociologist Ann Oakley argued that gender "is a matter of culture: it refers to the social classification into 'masculine' and 'feminine.' "[25] As John Hood-Williams suggests, Oakley's sex/gender distinction "enabled an oppositional stance to biologisms that attempted to tie women to subordinate positions on account of a largely immutable biology."[26] Defining sex as biological and gender as cultural, Oakley drew her conception of the sex/gender distinction directly from the work of psychoanalyst Robert Stoller and psychoendocrinologist John Money and his colleagues John and Joan Hampson in the endocrine clinic of the Johns Hopkins Hospital.[27] Summarizing their research, Oakley writes, "While Stoller talks about 'gender identity,' Money and the Hampsons refer to 'psychosexual orientation': the meaning of both terms is the sense an individual has of himself or herself as male or female, of belonging to one or other group. The development of this sense is essentially the same for both biologically normal and abnormal individuals, but the study of the biologically abnormal can tell us a great deal about the relative parts played by biology and social rearing: there are a multitude of ways in which it can illuminate the debate about the origin of sex differences."[28] Oakley's uncritical acceptance of the normal/abnormal distinction as a biological given reiterates a foundational epistemological presumption underlying Stoller's and Money and the Hampsons' research and biomedical research more generally: that humans may be naturally divided into clear and discernable normal and pathological types.[29] In recent years, scholars in feminist, queer, critical race, disability, postcolonial, transgender, and intersex studies have shown this presumption to be culturally and politically motivated.[30]

The pathologizing aspects of Oakley's account become particularly evident in the way she frames intersexuality. Analyzing several case studies from Stoller's work, Oakley argues that "parents' attitudes in rearing"[31] have a strong effect on children's gender presentations. She then turns to

Money and the Hampsons, suggesting that "[c]ase studies of individuals, though fascinating, cannot alone support sweeping generalizations about the lack of identity between sex and gender. A large group of hermaphroditic patients have been studied by Money and the Hampsons, and in 95% of all the cases (totaling 113, which is a large number for this sort of abnormality) *the sex of rearing corresponded to gender identity*. Most significantly, the correspondence held even for those individuals whose sex of rearing contradicted their biological sex as determined by chromosomes, hormones, gonads and the formations of the internal and external genitals."[32] Oakley used Money and the Hampsons' data on intersexual patients to forward a theory of gender's social construction. If gender is socially constructed, Oakley hypothesized, then gender roles and inequalities are changeable. Of course, more recent studies of intersexuality and gender roles, including those examining the highly publicized case of David Reimer, would call into question Money and the Hampsons' initial findings regarding the plasticity of gender.[33] From today's perspective, it is possible to see the leap in logic underlying both Money and the Hampsons' thesis and Oakley's feminist appropriation of it. The claim that gender is constructed is not reducible to the claim that rearing has a monocausal effect on gender presentation or identification. According to Vernon A. Rosario, biology and culture intertwine in complex ways in the formation of gender identity.[34] Contemporary work in feminist science studies, which I examine shortly, has also reached this conclusion.

Working within a nature/culture paradigm that presumed the two terms to be strictly oppositional, Oakley put pressure on the culture side of the equation to stress that gender roles, notably those that perpetuate male domination and female subordination, were learned, not inborn. "Sex differences may be 'natural,'" Oakley postulated, "but gender differences have their source in culture, not nature."[35] Putting the "natural" in quotation marks, Oakley contended, as would many feminists who followed in her footsteps, that social structures perpetuate gender inequalities by naturalizing them as innate sex differences. While extremely valuable as a critique of the workings of patriarchal power, this argument was founded on at least two assumptions that later feminists would call into question: first, that sex is purely biological; and second, that sex and gender are naturally and normatively dimorphic. Thus, even as she challenged the claim that gender roles reflect innate differences between the sexes, Oakley simultaneously consolidated a binary understanding of gender as the basis of a feminist politics of women's liberation. Concluding that "the aura of naturalness and inevitability that surrounds gender-differentiation in modern society comes, then, not from biological necessity but simply from the beliefs people hold about it,"[36] Oakley was unable to question the extent to which those beliefs are grounded in the presumption that binary ways of interpreting the world are natural and normal.

As some feminist theorists began arguing in the late 1980s and early 1990s, positing sex as the basis of gender fails to account for the sociocultural constitution of biological sex itself.[37] The feminist most cited for formulating this argument is Judith Butler, whose 1990 monograph *Gender Trouble* queried how the regulatory operations of what she called "the heterosexual matrix" maintain various sexual hierarchies.[38] Butler intervened in feminist debates over the sex/gender distinction by questioning the idea that biological sex is the foundation of cultural gender. Challenging the assumption that sex forms the natural substance onto which the social meaning of gender is written, Butler proposed that "gender ought not to be conceived merely as the cultural inscription of meaning on a pregiven sex (a juridical conception); gender must also designate the very apparatus of production whereby the sexes themselves are established."[39] In this view, gender is not simply a system of meanings imposed onto bodies but is rather an "apparatus of production," a generative technology that naturalizes the illusion of a prediscursive sex. If "sex itself is a gendered category," as Butler contends, then "sex" must be understood as a product rather than the cause or ground of gender.[40]

In arguing that gender produces the discursive and cultural notion of sex, Butler was also suggesting that gender should not be conceived as a substantive identity but rather as a process, a kind of ongoing doing, what she calls "a constituted *social temporality*."[41] Gender, Butler powerfully proposed, is performative in the sense that it is tenuously constituted by the very acts that are said to merely express it. In Butler's words, "the very notions of an essential sex and a true or abiding masculinity or femininity are also constituted as part of the strategy that conceals gender's performative character and the performative possibilities for proliferating gender configurations outside the restricting frames of masculinist domination and compulsory heterosexuality."[42] Butler's influential work contrasts with Oakley's precisely by being a prime example of a feminist project that contests the presumption of natural sexual dimorphism by using a poststructuralist framework to destabilize foundationalist accounts of nature.[43]

Since the publication of *Gender Trouble*, feminist scholars have problematized the theoretical underpinnings of the sex/gender distinction by attending to the ways in which sex and gender fail to neatly align with each other and with the nature/culture distinction within and across a variety of historical and contemporary contexts.[44] Butler is just one example of a feminist theorist who has productively troubled the presumed coherence and stability of the sex/gender distinction. More recent theorists, especially those working between feminist, queer, and transgender studies, including Jack Halberstam, Gayle Salamon, and Jean Bobby Noble, have used even as they have transformed prior feminist analyses of sexual and gendered hierarchies into occasions for the radical denaturalization of both gender and sex.[45] In this context, it seems important to note that the denaturalization effort would be impossible without the ongoing critique of the production of hierarchies based on sex and gender.

In recent years, there has also been what some have called a return to biology in feminist theory. Anne Fausto-Sterling, Elizabeth A. Wilson, Karen Barad, Deboleena Roy, and others have asked what happens when feminist theory goes beyond a critique of the sciences and takes biological material and scientific practice seriously.[46] Most pertinent to my analysis here is Fausto-Sterling's work. In *Sexing the Body*, Fausto-Sterling investigates how various scientific disciplines—endocrinology, genetics, neuroscience, and other fields—produce knowledge about gender, sex, and sexuality, and she argues against the dualisms of nature/culture, sex/gender, male/female, and heterosexuality/homosexuality. Fausto-Sterling suggests that accounts of embodiment cannot afford to discount biological processes, yet she simultaneously stresses that biological processes are not exterior to culture. Reading biological data through feminist theory to contest sexual dimorphism, Fausto-Sterling also implicitly foregrounds the regulatory character of gender as a system of power: "Our bodies are too complex to provide clear-cut answers about sexual difference. The more we look for a simple physical basis for 'sex,' the more it becomes clear that 'sex' is not a purely physical category. What bodily signals and functions we define as male or female come already entangled in our ideas about gender."[47] Suggesting that sex cannot be definitively disentangled from gender, Fausto-Sterling turns to the history of the medicalization of people with intersex to show that scientific research is influenced by culture, that both shape how bodies come to matter: "Intersexuals, seen as deviations from the norm who need to be 'fixed' in order to preserve a two-gender system, are also studied [by medical professionals] to prove how 'natural' the system is to begin with."[48] Fausto-Sterling's analysis of this contradiction, and of the medical and sociopolitical history of intersexuality more generally, demonstrates that science does not merely reflect but actually contributes to the production of cultural norms through its own supposedly value-free practices.

Fausto-Sterling and other feminist science studies scholars have offered important reconsiderations of entrenched epistemic paradigms in both the sciences and feminist theory. Moreover, their work challenges the mind/

body and nature/culture dualisms in ways that differ significantly from Butler's. Rather than privileging discursivity, they adopt a developmental systems theory approach that frames biology and culture as tied together in a multidimensional feedback loop. This innovative interdisciplinary approach reemphasizes the vital role that women's, gender, and sexuality studies can play in transforming the disciplinary and epistemic divides that structure the contemporary university.

The literature reviewed above illustrates some of the diverse ways feminists have rethought sex and gender. In the next section of this chapter, I suggest that feminist conceptualizations of the relation between sex and gender provide a critical basis for understanding how mid-twentieth-century medical specialists formulated what would become the dominant paradigm of intersex treatment and for critically rethinking the body politics of sex and gender normativities.

Gender in Money's Research

Lurking behind this feminist story is the figure of John Money. As the inventor of the term *gender role*, Money's work brings into focus the role of intersex as an origin of *gender* and of the sex/gender distinction. Indeed, as I will suggest, thirty-five years before *Gender Trouble*, Money posited gender as *prior* to sex.

Though the *Oxford English Dictionary* attributes the formulation of gender as a concept that emphasizes the social and the cultural to Oakley, the term *gender* actually began to crystallize as a category with a meaning distinct from biological sex in English at least twenty years earlier.[49] As Hausman and Germon observe, gender first emerged as an explicit object of inquiry in the behavioral and hard sciences in the mid-1950s, specifically in Money's psychobiological research.[50] Hausman argues that Money's research produced "a discourse about the body and human identity in sex that became powerful both as a justification for medical practices and as a generalized discourse available to the culture at large for identifying, describing, and regulating social behaviors."[51] That discourse hinged on a particular conceptualization of gender that played a key role in intersex medicine to justify surgical normalization and, more broadly, became a technology for regulating human behavior and embodiment writ large.

While studying the relation between endocrine functions and psychological states of hermaphroditism at Harvard in the 1950s, Money coined the term *gender role* as a diagnostic category and treatment protocol for patients whose anatomical configurations were regarded as unintelligible within the dominant frame of dimorphic sex. For people with intersex characteristics, whose bodies Money read as improperly sexed, *gender role* became a way for Money to predict and, as we will see, to literally fashion the sex they were "supposed" to have been all along. Money's typical scientific approach used the abnormal to find and define the normal. His work on intersex helped to popularize the view that gender is central to the sexual health of persons in general.

Money first made reference to his theory of gender in a 1955 article published in the *Bulletin of the Johns Hopkins Hospital* titled "Hermaphroditism, Gender and Precocity in Hyperadrenocorticism: Psychologic Findings."[52] In that paper, Money would later write in a 1995 essay, "the word *gender* made its first appearance in English as a human attribute, but it was not simply a synonym for *sex*. With specific reference to the genital birth defect of hermaphroditism, it signified the overall degree of masculinity and/or femininity that is privately experienced and publicly manifested in infancy, childhood, and adulthood, and that usually though not invariably correlates with the anatomy of the organs of procreation."[53]

This sentence is taken from a retrospective essay Money wrote on his life's work titled "Lexical History and Constructionist Ideology of Gender." It is the opening chapter of his collection of essays *Gendermaps*, where Money defends the science of psychosexual research against charges from feminists and social constructionists.[54] In his 1995 language, Money calls hermaphroditism a "genital birth defect," and this pathologizing rhetoric figures hermaphroditism as a problem of genital formation. However, in his earlier work Money clearly recognized the existence of a variety of intersex conditions that are irreducible to considerations of genital formation.[55] This reductionism reveals that what Morgan Holmes calls "genital determinism" came to play a significant role in Money's later thinking.[56]

In their influential textbook *Man & Woman, Boy & Girl*, Money and Anke E. Ehrhardt offer a more general theory of hermaphroditism, claiming that the terms hermaphroditism and intersex can be used interchangeably, as both "mean . . . that a baby is born with the sexual anatomy improperly differentiated. The baby is, in other words, sexually unfinished."[57] Two presuppositions ground this claim: first, that sexual anatomy has a proper mode of differentiation that, second, constitutes a complete or finished form of sexual dimorphism. In addition, in labeling intersex infants "sexually unfinished," Money and Ehrhardt reveal the persistence of a commonplace medico-scientific attitude toward abnormality analyzed by Michel Foucault in *Abnormal*.[58] In his genealogy of abnormality in western culture, Foucault observes that hermaphroditism played a special role in the formation of "the very first rudiments of a clinical approach to sexuality" in sixteenth-century Europe.[59] Between the sixteenth and seventeenth centuries, European medical men understood hermaphroditism within a juridical-natural model as monstrous. However, by the eighteenth century, doctors began to conceptualize hermaphroditism not as a breach of nature but rather as a defective structure. This view allowed European medical practitioners to articulate their role in regard to hermaphroditism as not simply diagnostic but as corrective or normalizing. In accordance with this view, Money and Ehrhardt's understanding of intersex was not only pathologizing but was also structured by a spatial and temporal logic of human development whose telos is wholeness. As several critics have pointed out, this perspective is problematic in terms of its heteronormative and sexually dimorphic ideological biases.[60] It is also fundamental to the logic of normalization Foucault discusses as emerging in the eighteenth century in *Abnormal*.

These presuppositions were evident in Money's work from the start. Money first became acquainted with hermaphroditism in the Harvard psychological clinic, where he wrote his PhD dissertation on "Hermaphroditism: An Inquiry into the Nature of a Human Paradox."[61] For his dissertation, Money conducted 10 case studies with interviews and collected 248 cases from a medical literature review to show that "psychosexual orientation bears a very strong relationship to teaching and the lessons of experience and should be conceived as a psychological phenomenon."[62] By "psychosexual orientation," Money meant "libidinal inclination, sexual outlook, and sexual behavior."[63] In "Lexical History and Constructionist Ideology of Gender," Money quotes his dissertation at length to reveal how his studies of hermaphroditism generated for him the following problem: "For the name of a single conceptual entity, there are too many words in the expression 'libidinal orientation, sexual outlook, and sexual behavior as masculine or feminine in both its general and its specifically erotic aspects.' The challenge to give a unitary name to the concept embodied in these many words became pressing after my case load of hermaphrodites studied in person had, after 1951, expanded from ten to sixty in Lawson Wilkins' Pediatric Endocrine Clinic at the Johns Hopkins Hospital, at which time a concise report of the findings became essential."[64] Studying individuals with anatomical configurations he regarded as anomalous, Money initially and inadvertently proliferated diagnostic categories; his research generated, he says, "too many words." This excess of signification highlights the degree to which intersexuality troubled the symbolic resources of Money's biomedical episteme. To overcome the discursive proliferation that his studies of intersexuality inaugurated, Money went in search of "a unitary name." In short, Money sought to establish an exhaustive, monolithic taxonomy to explain and contain the discursive excess generated by hermaphroditism. Money's project was to produce a coherent medical science of the abnormal along the lines discerned by Foucault.

Money's dissertation suggested that psychosexual orientation is shaped by social and psychological factors, and in forwarding this thesis Money was staging an argument with previous psychologists and sex researchers who held that psychosexual orientation was biological and innate. In the 1950s, a time when biological determinism, while contested, was still dominant in the hard sciences,[65] Money's insistence that masculinity and femininity could not be reduced to biology alone remains quite remarkable. Summarizing his post-1951 findings, Money explains in "Lexical History and Constructionist Ideology of Gender" that

> The first step was to abandon the unitary definition of sex as male or female, and to formulate a list of five prenatally determined variables of sex that hermaphroditic data had shown could be independent of one another, namely, chromosomal sex, gonadal sex, internal and external morphologic sex, and hormonal sex (prenatal and pubertal), to which was added a sixth postnatal determinant, the sex of assignment and rearing . . . The seventh place at the end of this list was *an unnamed blank that craved a name*. After several burnings of the midnight oil I arrived at the term, gender role, conceptualized jointly as private in imagery and ideation, and public in manifestation and expression.[66]

The "hermaphroditic data" led Money to the hypothesis that biological sex is itself radically unstable, composed of heterogeneous elements that do not add up to a unitary conceptual entity. Reckoning with this instability produced for Money a problem of language and reference, a problem of naming (earmarked by his peculiar tautology "an unnamed blank"). The "unnamed blank that craved a name" to which Money refers in this passage can be read as a displacement of the biological instability exposed by intersexuality. In other words, in recognizing a list of prenatally and postnatally "determined variables of sex that hermaphroditic data had shown could be independent of one another," Money's research dismantled the unitary conception of sex and, in so doing, produced an "unnamed blank" at the site of the body. This "unnamed blank" threatened the very semblance of sex. To contain that threat, Money filled the blank with gender. Put differently, Money used gender role to name and thereby semantically fill (or cover over) the void left by sex's lack of conceptual and referential unity.

As Germon puts it, "at a pragmatic level, gender provided a solution to the uncertainty of any absolute somatic sex. Gender served to stabilize what advances in medical technology had rendered more and more unstable during the first half of the twentieth century."[67]

While *gender role* offered stability where technology's destabilization of sex was concerned, it also gave Money a linguistic means to contain the discursive proliferation ("too many words") occasioned by his research on intersex. By giving the "unitary name" *gender role* to the "unnamed blank," Money introduced a seemingly coherent sign where he previously had found only unstable, discontinuous elements. Moreover, Money anthropomorphizes the "unnamed blank"—he attributes to it the "craving" for "a name"—making it seem as if the unnamed blank were itself a subject of desire, longing for epistemic certainty and representational unity, yearning, in short, for someone to give it a name. Giving the "unnamed blank" the name *gender role*, Money proceeds as if that naming could guarantee a relation of referential coherence between word and inchoate object. This anthropomorphism dissimulates Money's own medico-scientific craving for epistemic positivity. By figuring gender role as the name craved by the unnamed blank, Money thus overrides and conceals intersexuality's undoing of the structure and stability of sexual dimorphism and makes the internal and external manifestation of masculinity or femininity the pinnacle of his classificatory schema.

In defining *gender role* in terms of interior and exterior expression as masculine or feminine, Money was extrapolating from what feminist political theorists would later argue is a gendered political construction through and through: the public/private distinction that emerged in the western world in eighteenth-century social contract theory.[68] Treating this uninterrogated public/private distinction as an abiding feature of gendered subjectivity, Money recognized that gender role's unity was not a given. One could privately identify as feminine yet publicly manifest a masculine identity, or vice versa. This apparent contradiction suggested to Money that gender role was acculturated and imprinted at multiple levels of a person's psychosexual orientation and that these levels were not automatically coherent with one another. Money believed that psychoendocrinology could resolve this potential incoherence. Medical technology, he posited, could produce what nature could not guarantee: the unity of interior and exterior expressions of gender.

This helps to explain why Money approached "gender role," as he says in the above passage, as a "variable of sex." That is, though Money disaggregated gender role from sex, he also posited a structural connection between them. As the term *variable* indicates, gender role signified for

Money not only "masculine or feminine inclination, outlook, and behavior," but also the prospective sex that is supposed to coincide with a particular gender role. In this way, Money posited *gender role* as a predictive agent to determine the hermaphrodite's sex. In short, long before Butler, Money proposed that gender precedes sex.

In contemporary feminist theory, the postulation of gender as prior to sex has been a touchstone for antifoundationalist accounts of embodiment. For Butler, for instance, the reversal of the conceptual polarity of the sex/gender distinction represents the first subversive gesture in a two-pronged deconstructive movement of reversal and displacement. But it is crucial to recognize that Money's superordination of gender over sex was not a subversive gesture but rather a regulatory one. By determining a hermaphroditic infant's prospective gender role, Money was then retroactively able to determine the infant's sex as male or female, and this is why his treatment recommendations centered on surgical, hormonal, and psychosocial normalization. In using "gender role" to fill the "unnamed blank" intersexuality represented, Money attempted to make individuals born with intersex characteristics fit into normative schematizations of the roles conventionally embodied by people with dimorphic sex.

As Iain Morland notes in "Cybernetic Sexology," Money claimed to think "cybernetically" about sex and gender, and his usage of the word *variable* in this instance is a prime example. However, according to Morland, Money made a formative error "in his application of cybernetics to sexology. Cybernetics theorized dynamic systems that can adapt, not merely repeat. It was therefore irreconcilable with the sudden, irrevocable establishment of gender in infancy that was axiomatic for Money."[69] At the very moment when his research pointed toward potentially radical instabilities between gender and sex—and within gender and sex themselves—Money erased those possibilities by reducing gender to the performance of the roles he thought dimorphic sex *should* entail—that is, by fixing gender as mere repetition, as axiomatic, to use Morland's terms. As Hausman points out, what Money "argued, in effect, was that those subjects unable to represent a sex 'authentically' could simulate one through adequate performances of gender that would fix one's identity irrevocably in a sex category. In other words, if you aren't born into a sex, you can always become one through being a gender."[70] Though I agree with Hausman that Money used gender to restabilize sex, my analysis diverges from hers on the question of gender's so-called authenticity.

For Hausman, "the idea of gender" is a discursive construction of psychiatry. Hausman further suggests that gendered interiority is a product of technology and discourse and is therefore artificial.[71] In his critical review of *Changing Sex*, Rosario rightly challenges Hausman on this point.[72] Rosario argues that Hausman ultimately "relies on a rigid internalist, technological-determinist historiography."[73] Rosario further contends that "it is hard to give full credit to Money for inventing gender identity when late-nineteenth-century doctors, such as the Italian forensics expert Arrigo Tamassia, clearly defined the conflict between psychological gender identity and physical sex appearance in certain cases of 'sexual inversion': 'the individual, although recognizing himself of a given sex, psychologically feels all the attributes of the opposite sex.' (Tamassia, of course, like the Italians and French of today, lacked a linguistic means of making the current, English 'sex'/'gender' distinction)."[73] In Rosario's view, the category gender need not explicitly exist as such in a particular culture's language for the sex/gender distinction to be operative in that culture. While I would concede the plausibility of this point and willingly acknowledge that *gender* can be traced back to multiple points of origin, attention to the particular conceptualization of gender advanced by Money and adopted by Oakley nonetheless highlights a crucial linkage between mid-twentieth-century biomedical and feminist discourses that Rosario leaves unremarked.

As Morland notes in his introduction to the 2009 special issue of *GLQ* "Intersex and After," the role of gender in the development of intersex treatment, and in Money's research in particular, remains contentious.[75] In the paragraph from his 1955 article "Hermaphroditism, Gender, and Precocity in Hyperadrenocorticism," in which the term first appeared, Money theorized *gender role* as pertaining specifically to the way in which behavior cannot be causally linked to biological sex: "Cases of contradiction between gonadal sex and sex of rearing are tabulated . . . together with data on endogenous hormonal sex and gender role. The term gender role is used to signify all those things that person says or does to disclose himself or herself as having the status of boy or man, girl or woman, respectively. It includes, but is not restricted to sexuality in the sense of eroticism."[76] Money then offered the following summary conclusion, which I quote at length:

> Chromosomal, gonadal, hormonal, and assigned sex, each of them interlinked, have all come under review as indices which may be used to predict an hermaphroditic person's gender—his or her outlook, demeanor, and orientation. Of the four, assigned sex stands up as the best indicator. Apparently, a person's gender role as boy or girl, man or woman, is built up cumulatively through the life experiences he [*sic*] encounters and through the life experiences he [*sic*] transacts. Gender role may be likened to a native language. Once ingrained, a

> person's native language may fall into disuse and be supplanted by another, but it is never entirely eradicated. So also a gender role may be changed or, resembling native bilingualism may be ambiguous, but it may also become so deeply ingrained that not even flagrant contradictions of body functioning and morphology may displace it.[77]

Historically linked with the concepts of ethnicity and the mother tongue, a native language is the first language one learns. It is learned, but almost as soon as it is learned, it becomes habitual, reflexive, deeply rooted, *ingrained*, to use Money's term, almost as if it were natural. Ingrained means "firmly fixed," but the term also refers to a type of yarn, dyed before weaving, used to make reversible carpets. Analogizing gender with a native language and emphasizing gender's ingrainability, Money figures gender as both text and textile. Texts are of course literally stitched of fibers. But every textile is also a social text.[78] Money's analogy positions gender as dynamic object. Formed over time, related to but not fully determined by both biology and culture, changeable under certain conditions but not always intentionally so, gender's shape, stability, and permanence have no guarantees. Yet Money implies that western biomedicine can comprehend this object through scientific study and thereby attempt to mold it.

In analogizing gender with text and textile, Money was not only contemplating gender's moldability but also simultaneously prefiguring and effacing one of the lessons of poststructuralist feminisms: that gender is structured like a language, a system of differences without positive terms.[79] If gender is like a language, then gender is not only a relational system but also a system where the meaning of any given term is both arbitrarily and negatively determined. But Money forecloses this insight by positing the existence of proper, positive binary terms as the ground of the system: "*his* or *her* outlook, demeanor, and orientation" (emphasis added). Money's normative dimorphic prerogative and his investment in the propriety of binary logics come together to privilege heteronormative masculine and feminine roles and bodies as regulatory ideals, over and above alternative possibilities of comportment, identification, and embodiment.

Money's reference to "native bilingualism" as "ambiguous" is also noteworthy. The figure marks native bilingualism as indefinite, unclear, and confusing, when in fact native bilingualism just means that a person grows up speaking two languages. Bilingualism opens up opportunities for translation, raises questions about linguistic and cultural difference, and reveals the promise of border crossing. It destabilizes those nations and cultural traditions that privilege the idioms of monolingualism and ethnocentrism.[80] In other words, Money's position on native bilingualism is westocentric in terms of its assumptions about the nature and normativity of racial, ethnic, and class hierarchies. He views bilingualism not merely as a threat to gender but also as a threat to the unmarked whiteness of normative gender roles. Money codes categories and bodily configurations that trouble expected boundaries and forms, disrupt cultural norms and preconceptions, and challenge ideas of sovereignty and wholeness as a threat to intelligibility. As with the "unnamed blank" analyzed above, Money's diagnostic effort becomes regulatory in multiple ways, an effort to contain embodied subjects whose differences generate ambiguities and proliferate languages, cultures, and meanings.

As Downing, Morland, and Sullivan note, Money's understanding of bilingualism shifted over the course of his career. During the 1950s and 1960s, Money used bilingualism as a figure for improper gender development, as evidenced by the above passage. However, from the 1970s on, "Money presented bilingualism as akin to *proper* gender development" and argued that an ordinary child's exposure to two languages was like her or his exposure to two genders, one of which she or he will imitate.[81] Acknowledging the flawed and reductive logic of Money's analogy between bilingualism and gender role acquisition, Downing, Morland, and Sullivan emphasize that even as Money's understanding of bilingualism changed over time, in each case he sought to reduce "bilingualism" "to a single, unequivocal phenomenon."[82]

The regulatory aspect of Money's work is especially apparent in the gendered language that shapes the passage I have been reading. Between the first and third sentence, there is a grammatical shift from *his or her* to *he*. The first sentence reads, "Chromosomal, gonadal, hormonal, and assigned sex, each of them interlinked, have all come under review as indices which may be used to predict an hermaphroditic person's gender—*his* or *her* outlook, demeanor, and orientation" (emphasis added). The third sentence reads: "Apparently, a person's gender role as boy or girl, man or woman, is built up cumulatively through the life experiences *he* encounters and through the life experiences *he* transacts" (emphasis added). Here, Money switches to the masculine singular pronoun, using it as the general form of personhood. This usage reveals the masculinism, or, more precisely, the masculine universalism that guides Money's project, a masculine universalism evident not only at the level of grammar but also in the conceptual transition from hermaphroditism to binary gender. Money resolves the tension between the destabilization and multiplication of sexes and sexed subject positions inaugurated by his research on intersexuality and binary grammar by privileging the masculine singular pronoun as the signifier of universal personhood.

This masculinism was central to Money's deployment of gender role as a category of prediction: "Chromosomal, gonadal, hormonal, and assigned sex, each of them interlinked, have all come under review as indices which may be used to *predict* an hermaphroditic person's gender—his or her outlook, demeanor, and orientation" (emphasis added). By disaggregating gender role from biological sex, Money was able to interpellate an intersex person's psychosexual orientation in terms of dominant conceptions of masculinity and femininity regardless of the individual's morphological "sex." This disaggregation also provided Money with a paradigm of treatment. By the mid-twentieth century, the discipline of surgery had advanced to the point where doctors could perform surgical sex reassignments.[83] Yet sex reassignment could only be framed as medically necessary if it could be shown that an intersex infant's psychosocial orientation could be predicted. Money's theory of gender role filled precisely that gap.

Money theorized sex as surgically malleable and gender as socially plastic to maintain the binary order of things. As Morland observes in "Gender, Genitals, and the Meaning of Being Human," "the notion of human genitals and gender as surgically and socially plastic depends on the conceptualization in twentieth-century science of plasticity as a quintessential human attribute."[84] Money conceptualized, Morland continues, "genitals and gender as malleable at a historical moment when plasticity and humanity were held by Western science to be equivalent. This had the mutually reinforcing effects of facilitating the uptake of Money's ideas about how to treat intersex, while instituting gender as a core human quality, flexible by definition."[85] In other words, Money's use of gender as a predictive agent presumed that humans are plastic enough to tolerate treatment in the first place.

In devising a course of treatment for intersexuality, Money, along with fellow researchers at the Johns Hopkins Psychohormonal Research Unit, formulated what has come to be known as the optimal gender paradigm. They held that "the sex of assignment and rearing is consistently and conspicuously a more reliable prognosticator of a hermaphrodite's gender role and orientation than is the chromosomal sex, the gonadal sex, the hormonal sex, the accessory internal reproductive morphology, or the ambiguous morphology of the external genitalia."[86] As Rosario explains, the Hopkins team "argued that infants born with ambiguous genitalia could be surgically 'corrected' and then successfully raised as either males or females so long as certain conditions were met."[87] These conditions included gender assignment before eighteen–twenty-four months; that parents strictly enforced the gender of rearing; and that the children were "not confused by knowledge about their intersexed past."[88] According to Alice Dreger and April Herndon, the optimal gender paradigm "held that *all* sexually ambiguous children should—indeed must—be made into unambiguous-looking boys or girls to ensure unambiguous gender identities."[89] In other words, if gender is like language, and gender instability (changing genders) is like native bilingualism, Money's ultimate goal was to eradicate ambiguity in the name of promoting monolingualism. This seems to resolve the problems of both discursive excess ("too many words") and linguistic inadequacy ("an unnamed blank that craved a name").

In recommending that intersex infants be treated with a combination of normalizing genital surgeries, hormonal treatments, and psychosocial rearing into the "optimal gender," Money and his colleagues essentially designed a program of sex *and* gender normalization premised on implicitly white and middle-class ideals of active masculinity and passive femininity. This program of normalization can also be understood as a refinement of the westocentric masculinism (disguised as grammatical) inherent in Money's privileging of the masculine pronoun. As Karkazis points out, Money and other intersex medical specialists' intentions were, to some degree at least, beneficent: "Raising a child with a gender-atypical anatomy (read as gender ambiguity) is almost universally seen as untenable in North America: anguished parents and physicians have considered it essential to assign the infant definitively as male or female and to minimize any discordance between somatic traits and gender assignment."[90] Money and the Hopkins team thought that their treatment protocols would help intersex children to live "normal" lives. Intersex activists and scholars have criticized these protocols, however, for inflicting physical and psychological trauma and upholding an unjust system of bodily and psychical regulation.[91]

Racializing/Queering Money's Gender Concept

From an intersectional and transnational perspective, Money's optimal gender paradigm must be understood as not solely an effort to restabilize biological sex, but also as a biopolitical and geopolitical project overdetermined by historically sedimented and naturalized western hierarchies of race, class, nation, and other categories of difference. In a forthcoming essay titled "Gone, Missing: Queering and Racializing Absence in Trans & Intersex Archives," Hilary Malatino argues that intersex and trans health care has been shaped by racialized and classed ideals and mores of sexual dimorphism and gender binarism—ideals and mores that were produced by settler colonialism, racial capitalism, and their systems of imperial knowledge production.[92] Investigating Money's and Alfred Kinsey's archives at Indiana University, Malatino offers a brilliant reading of the raced and classed absences that haunt intersex and trans medical archives. According to Malatino, these absences are not accidental but are rather strategies of deliberate disappearance from and, in some cases, active resistance to the imposition of what María Lugones calls the colonial/modern gender system.[93]

Lugones contends that the colonial/modern gender system found its rationale and justification not only in western structures of political, economic, literary, artistic, and cultural dominance that privileged the white, bourgeois, patriarchal family as the telos of civilized advancement, but also in the eugenic and sexological sciences. In the name of modernization and progress, eugenics—the scientific selection of so-called superior human genetic types—and sexology—the biosciences of sex—created idealized typologies of sexual dimorphism and binary gender that were dependent on and inseparable from acculturated racialized and classed ideologies of human development. "Lugones," Malatino explains, "reasons that white bourgeois ideals of gender embodiment have been shaped by a deeply dimorphic understanding of gender complementarity that emphasizes white female sexual submissivity, domesticity, minimized agency and access to the public sphere and white male providence, epistemic and political authority, virility, and naturalized dominance. This 'light side' [Lugones's term] of the colonial/modern gender system stands in contrast to [what Lugones calls] a 'dark side,' constituted by the ways in which the sexualities, embodiments, and kinship forms of colonized peoples were constructed within the colonial imaginary" as inferior, deviant, and pathological.[94]

Malatino's reading of Money through Lugones's decolonial feminist lens provides an invaluable rethinking of the co-constitution of race, class, nation, *and* gender in genealogies of intersex. As Malatino discovered,

> trans and intersex folks of color are conspicuously missing from the medical archives of sexology; moreover, many folks—white folks and folks of color—appear briefly in medical records, only to never return, in effect going AWOL from the medical protocols of transition and gender normalization. Despite this, never in the work or correspondence of either of these massively influential sexologists [Kinsey and Money] I've researched was there any reflection on the partiality of knowledge manifest in such a racially homogenous, westocentric archive. In the rare moments that folks of color appear in these archives, they are

> framed, in accordance with the logic of the "dark side" of the colonial/modern gender system, as deviant, sexually perverse, and culturally both aberrant and anachronistic.[95]

Malatino's analysis helps to explain the unmarked whiteness of Money's heteronormative, masculinist approach to intersex normalization. Her work also highlights the ways in which intersex people of color occupy a crucial place in the medical archives of sexology precisely because they are simultaneously "conspicuously absent" from even as they sometimes briefly appear in those archives, only to subsequently disappear by refusing to show up for their follow-up visits with their doctors. Malatino may be slightly overstating the extent of this conspicuous absence. Limited data as well as anecdotal evidence suggest that some intersex people of color were medically normalized under Money's optimal gender paradigm during the latter half of the twentieth century, as I discuss in chapter 5. Nonetheless, Malatino's larger point—that structures of race and class dispossession limit access to medical treatment—helps us to grasp the intersectional disparities that unevenly shaped and continue to shape the medicalization of intersex, trans, and gender nonconforming folks.

The transient racialized and classed appearances/disappearances that Malatino documents have an especially queer valence to them, a queerness that resists precisely the metaphysics of presence that governs medical archives as artifacts of western hegemony and its contestations. Atypical sex and gender nonconformity could only be read as a quintessential threat to the intelligibility of normative whiteness in a context where people of color and poor people were already considered less than human. These haunting subjects thereby disseminate an underanalyzed form of what we might call, after Avery Gordon, ghostly agency.[96] As an*other* kind of "unnamed blank"—one that Money himself could not imagine or anticipate—these nonpresent racialized, classed, and queer figurations trouble and disorient the colonial/modern gender system by exposing its constitutive exclusions, asymmetries, and contradictions. As hauntological or necropolitical figures,[97] the subjects who occupy these unnamed blanks are of heterogeneous provenance. As such, their transience in the medical archives can perhaps only register as an indecipherable but insistent subaltern refusal of medico-scientific regimes of legibility and intelligibility.[98] Any attempt to translate this subaltern refusal into a liberal humanist version of agency will likely reiterate the violence of imperialism and settler colonialism.[99] Nonetheless, we might still ask: how can we begin to learn to responsibly respond to the histories of racialized, classed, and queer dispossession and disappearance disclosed and effaced in sexology's archives and biomedical practices?

For Malatino, any answer to this question must take care to locate the stakeholders of gender regulation in the subjects who embody its deepest margins. For this reason, Malatino concludes that

> racialized, classed, and queer absences are central to understanding how access to technologies of transition have become intensively compromised for poor folks, trans folks of color and gender non-conforming, non-heterosexual folks while they have, simultaneously, been coercively imposed on intersex folks in the interest of normalizing our divergent forms of sexed embodiment. This selective utilization of technologies of transition helps explain why it is we currently lack holistic approaches to trans and intersex health that move beyond surgical and hormonal techniques of gender normalization and focus more heavily on the structural violence that too often compromises the life chances of trans and intersex subjects.[100]

Building on Malatino, I want to suggest that the ethicopolitical critique of the medical management of intersex can be strengthened by supplementing it with the genealogical critique of gender. Whereas the ethicopolitical critique underscores medicine's failure to uphold the Hippocratic oath ("first do no harm"), the genealogical critique foregrounds linkages between the medical management of intersex and the biopolitical regulation of embodied *differences* more broadly. At the level of activist strategizing, making these connections visible could expose the limitations of an exclusive focus on patient-centered health care and its focus on the patient as a neoliberal consumer (as I discuss in greater detail in the conclusion to this book) and might thereby enable the articulation of a more expansive vision of sexual and gender justice for a multicultural, transnational world.

One activist group that has taken some steps in this direction is Organisation Intersex International (OII), which has board members in twenty countries across the globe. OII recognizes that the medicalization of intersex is inseparable from the colonial pathologization of the embodiments and kinship systems of people of color, the medicalization of transgender and disability, the stigmatization of queerness, and other forms of corporeal normalization. Responding to medical professionals' and some activists' recent embrace of the term *DSD* (disorders of sex development) in place of *intersex*, members of OII have created posters, t-shirts, and broadsides that contest the DSD nomenclature. One declares, "Sorry, We're Not Disordered," while another contains a "Warning" sign and skull and crossbones placed next to text that reads: "DSD: Death to Sex Differences. DSD = Eugenics, DSD = Heterosexism, DSD = Transphobia, DSD = Homophobia."[101] OII's intersectional critique of the DSD nomenclature frames the medicalization of intersex as fundamentally linked with multiple, overlapping systems of power-knowledge. Countering medicalization, OII adopts a human rights approach to intersex and promotes the values of anatomical diversity, bodily integrity, and informed consent.

If the ethicopolitical critique stresses the human rights of intersex individuals, the genealogical critique adds a historical dimension to this emphasis. It shows how deeply invested—and deeply troubled—the medical model has been in its quest to contain anatomical variation and the threat such variation poses to the epistemic and somatic certainty of dimorphism as a ground of western exceptionalism and hegemony. Recalling with Foucault that there is no outside to power relations, and recognizing that gender is not merely a repressive technology but also a productive one, the challenge is to imagine strategies for expressing human potential in ways that can embrace uncertainty and enhance practices of freedom without consolidating the psychosomatic harms of gender regulation.[102]

Conclusion

Through Money's work, gender became one of the cornerstones of the modern medical management of intersex. In *Gender Trouble*, Butler observes that "the mark of gender appears to 'qualify' bodies as human bodies; the moment in which an infant becomes humanized is when the question, 'is it a boy or girl?' is answered. Those bodily figures who do not fit into either gender fall outside the human, indeed, constitute the domain of the dehumanized and the abject against which the human itself is constituted."[103] Seen in this light, Money's project essentially concerned the humanization of people with intersex and unwittingly revealed how dehumanizing humanism can be for those born with anatomies that do not conform to the mythical norm.[104] Though Money's work has been questioned in recent years, many clinicians continue to follow his guidelines, viewing intersex infants as corporeally unintelligible at the moment of birth, only to immediately transport them into intelligibility through surgical, medical, and psychosocial normalization. As my analysis has shown, these bodily interventions follow the strict, masculinist-as-universalizing, binary constraints of a westocentric cultural grammar. Most parents and doctors are so overly invested in the question "Is it a boy or girl?" that they cannot imagine a world of other possibilities.

To begin imagining those possibilities, it seems imperative to acknowledge that, despite significant differences in epistemological and sociopolitical orientation, aim, and method, contemporary feminist and biomedical discourses continue to share an investment in the presumption of gender's plasticity. While second-wave feminists theorized the social construction of gender to critique the determinist fallacy that "anatomy is destiny," feminist scholarship of the last two decades has pushed at the limits of gender constructionism, asking whether the very frame of binary gender discloses and effaces the centrality of normative whiteness, middle-class respectability, compulsory able-bodiedness, and empire to the workings of sexism, heteronormativity, transphobia, and intersexism. In the process, the lines between feminist, queer, critical ethnic, disability, transgender, and intersex studies have become productively blurry or plastic themselves. Meanwhile, since the 1950s, biomedical experts in intersex treatment have sought to rewrite the destiny of anatomy, using surgery, endocrinology, and psychosocial counseling to compel people with "atypical" bodies to conform to the regulatory codes of the colonial/modern gender system. Paradoxically, then, Money and the physicians who came to widely embrace his paradigm during the twentieth century used the presumed plasticity of psychosocial gender and the surgical malleability of anatomical sex to reinforce the very ideologies that feminist thinkers of diverse stripes have attempted to contest by theorizing the plasticity and malleability—as well as the intersectionality—of gender, sex, sexuality, race, class, nation, and ability.

When Oakley appropriated gender from Money to articulate a feminist project that would liberate women from biological determinism, she set into motion a historical process whose ramifications continue to reverberate in feminist discourse and practice today. Considering her political aims, it is perhaps understandable that Oakley reduced gender to its most binary (and implicitly white) formulation, but the costs of this reduction were considerable. Intersexuality's place in the invention of gender has been largely erased from feminism's historical archives. While feminists have thoroughly demonstrated that gender role, gender identity, and gender as a system of power are neither equivalent with one another nor reducible to a common denominator,[105] their different meanings remain contingent not only on processes of racial, sexual, abled-bodied, and national formation but *also* on the medical pathologization of intersexuality.

The omission of intersexuality's role in the development of the gender concept from the historical narrative of contemporary feminism is problematic because it conceals a profoundly unsettling paradox. On the one hand, the concept of gender enabled feminists from the 1970s onward to bolster critiques of biologism and essentialism, refine oppositional analyses of patriarchal systems, and underline the social and political foundations of inequalities between women and men. On the other hand, the term and concept of gender was born out of a regulatory, masculinist, and westocentric project—the medical management of intersexuality—that masked itself through the humanist language of the betterment of all peoples. In short, when the concept of gender as social is perceived as liberatory, an important aspect of the concept's history is lost.

In light of this history, it is no longer feasible to assume that gender is in and of itself an obvious "common sense" good; a universal characteristic of the human species (even if a culturally and historically variable one); or an aspect of self that forms prior to and independently of the biopolitical and geopolitical relations of power that condition its intelligibility. Registering these paradoxes entails grappling with gender's incoherence, polysemy, and uneven effects. Simultaneously a vector of identification *and* disidentification, inclusion *and* exclusion, empowerment *and* coercion, subjectification *and* subjection, gender has many different gatekeepers and stakeholders, and the stakes of gender, various as they are, are rooted in the politics of difference.

This is precisely why a genealogical approach to the messy relations between bodies and the words and practices that name them is so important. As I have argued here, intersex troubles gender not only conceptually but also genealogically. A critical attention to the history of intersex disrupts and displaces gender's presumed coherence and meaning; reveals that gender cannot be reduced to a transhistorical given or a purely descriptive category; calls attention to the power relations that transect the lives of people whose bodies have been marked as gender's constitutive outside; and underscores the historical processes, antagonisms, complicities, and exclusions that have shaped the development of gender as a concept, object of knowledge, paradigm of sociality, and technology of subject formation. Thus, far from being a marginal topic, subfield, or esoteric specialization within women's, gender, and sexuality studies, intersex is actually central to the history of the analytic categories that have fundamentally shaped the diverse intellectual trajectories, paradigms, pedagogies, and politics of the field. The medicalization of intersex literally gave birth to gender.

Ultimately, if the medicalization of intersex constitutes a condition of the everyday biopolitics of modern western gender regimes and their globalizing circuits, then to become gendered as a subject is to be called to learn to ethically respond to a larger, uneven intersectional and transnational history of dispossession and disappearance—a history, to be sure, that was never gender's alone.

2

Intersex Trouble in Feminist Studies

> The ethical way to treat intersexed individuals is to preserve, rather than surgically abolish, the uncertainties their bodies provoke.
>
> —Iain Morland[1]

A colleague recently told me the following story. She was doing a first-round interview for an assistant professor position in women's, gender, and sexuality studies (WGSS) at a well-known small liberal arts college. After she answered the standard questions about her dissertation, research interests, and pedagogical approach, a member of the search committee asked her what she thought would be the most significant areas of study in WGSS in the near future. My colleague named intersectional feminisms, queer of color critique, new materialism, and intersex and transgender studies. Before she had a chance to explain why she thought these fields were important, a senior faculty member on the committee interrupted her. "Intersex and transgender," the committee member said with a huff, rolling her eyes. "Soon, we won't even be allowed to talk about real women anymore!" The other committee members became visibly uncomfortable. One laughed half-heartedly while the other two went silent. They all then looked to my colleague to anticipate her response. She stated, "*Real women*, perhaps we should talk about what you mean by that term." As my colleague later told me, to her this was no laughing matter at all.

This anecdote highlights the manifest unease some people feel in relation to the rise of interest in intersex and transgender studies in academic feminist contexts. It also reveals the political and intellectual stakes underlying that discomfort. Those stakes center in large part on the place of women in a field formation that was, after all, originally named *women's* studies. That is, they have to do with the various forms of epistemic, cultural, and political privilege that accrue to the sign *women*—not only as the field's proper object of study, but also as the field's *raison d'être*.[2]

As an epistemological, pedagogical, and political economic institution, WGSS is, as Rachel Lee has argued, a site of regulation.[3] The pressure to police the field's boundaries, to maintain a certain feminist status quo, comes from multiple sources within the field and beyond.[4] For the committee member who responded to my colleague's mention of intersex and transgender studies with an ostensibly playful joke that also expressed a micro-aggression, it is as if the rise of interest in intersex and trans issues in WGSS not only derails the field's focus on studying women's lives—as if trans and intersex women are not *real* women—but also threatens, in an ironic twist of fate, to institute a moratorium on the usage of the word *women* in the context of WGSS itself.

This chapter offers a response to that position. I seek to show that the rise of interest in intersex in feminist studies does not entail any such moratorium, though I do suggest that the trouble intersex occasions in the field deserves greater attention because it exposes particular uncertainties about embodiment that haunt dominant, residual, and emergent modes of knowledge production about bodies within feminist theory, the academy, and beyond. In chapter 1, I argued that John Money used "gender" to fill the "unnamed blank" of intersexuality. Money's "unnamed blank that craved a name" is in many ways a story about the anxieties surrounding uncertainties. In this chapter, I suggest that the critical tools of feminist and queer studies make a different approach to intersexuality possible, an approach that allows for the ethical preservation of what Iain Morland calls the uncertainties intersex bodies provoke.

Analyzing foundational works by three scholars who have indelibly shaped contemporary feminist and queer scholarship—Suzanne J. Kessler, Anne Fausto-Sterling, and Judith Butler—I argue that intersex does not merely represent an expansion of the range of objects studied in feminist and queer work, but rather critically reconfigures key interdisciplinary analytic concepts: sex, gender, sexuality, embodiment, and the politics of difference more broadly. I contend that Kessler, Fausto-Sterling, and Butler have productively interrogated and reformulated the sex/gender, nature/culture, and gender/sexuality distinctions, interrupting heteronormativity's equation of genitalia with sex and with gender, thereby refiguring the matter of bodies as inextricable from biopolitical processes that position human anatomy and morphology as objects of regulation. In addition, I address the problems of intersex exceptionalism and the fetishization of intersex alterity as they play out in Kessler, Fausto-Sterling, and Butler and reflect on the role these problems continue to play in contemporary intersex studies.

Intersex and the Medical Construction of Gender

The current interest in intersex in feminist studies arises, in part, from the ways in which the topic is seen to generate key questions about the meaning and materiality of sex, gender, and sexuality. As I demonstrated in the introduction, the definition of "intersex" is highly contested. In a transnational world permeated and organized by strict yet also divergent and nonidenticial ideas about sex and gender, intersex phenomena raise important questions in relation to widely held assumptions about the nature of embodied difference. What defines a woman, what defines a man, and how do we know? How many sexes are there? Are male and female the only options? What systems delimit those options and how? These questions, arguably originary analytic points of departure for feminist studies, are of inestimable value for investigating the relations of power and knowledge that inform gendered and sexed biological, social, political, and subjective formations. When an infant is born with anatomical features that challenge accepted standards of sex/gender classification, with genitals, for instance, that are not readily readable in strictly dimorphic terms, naturalized presumptions about the ontology and epistemology of embodiment threaten to come undone. Entrenched gendered and sexed ways of being, knowing, and acting in the world are normative; they perform and participate in disciplinary operations that simultaneously produce and foreclose particular forms of life. As Judith Butler argues, the undoing of restrictive conceptions of sexed and gendered existence is a vital but tenuous prospect.[5] As much as the destabilization of sexed and gendered identities, meanings, and structures holds the potential to expand the range of possibilities for what gets to count as an intelligible, valuable life, so too does it harness the power to throw certain lives into crisis or render them inhabitable only under certain highly constricted conditions.

As I detailed in chapter 1, much feminist discourse from the early 1970s through the present has adopted Oakley's influential sex/gender distinction, where gender is understood as the set of sociocultural meanings, roles, and behaviors codified along the axis of masculinity and femininity that are mapped onto biological sex.[6] By contrast, feminist engagements with intersex have problematized the assumptions that give the sex/gender distinction its analytic purchase and motility.[7] Suzanne J. Kessler was one of the first scholars to address the issue from a feminist perspective and was also one of the first to put intersex on the agendas of feminist and queer research. In her 1990 *Signs* article titled "The Medical Construction of Gender: Case Management of Intersexed Infants," which she later expanded into her 1998 book *Lessons of the Intersexed*, Kessler critically examined the medical treatment of infants born with intersex features or "conditions."[8] As Kessler discovered researching the medical literature on intersexuality, most though not all intersex conditions pose little to no physiological health risk to the life of the infant.[9] This observation led Kessler to ask why health care experts view the birth of an infant with an intersex condition as a psychosocial "emergency," one requiring swift surgical intervention. In Kessler's view, the medical management of intersex infants set into motion a coordinated effort of "crisis management," an effort aimed at containing the epistemological and ontological anxieties and uncertainties that arise when dominant presumptions about the meaning and materiality of sex and gender begin to display their contingency, precariousness, and instability.

According to Kessler, the medical management of intersex infants literalizes in a striking fashion an argument central to the history, practice, and theory of feminisms: that gender, contrary to popular belief, is made, not given. Analyzing the medical literature on intersexuality alongside material from interviews she conducted with surgeons and endocrinologists, Kessler argues that "medical teams have standard practices for managing intersexuality that rely ultimately on cultural understandings of gender."[10] The key phrase in Kessler's thesis is "cultural understandings of gender." It suggests that frameworks for making sense of gender have their basis in culture, an argument that, through the work of scholars like Donna Haraway, Judith Butler, Diana Fuss, Diane Elam, and others, has become increasingly popular in feminist thinking since the early 1990s.[11]

Unlike many of her contemporaries in feminist theory and activism, however, Kessler did not take the cultural construction of gender to be analogous to what Gayle Rubin named in 1975 "the set of arrangements by which a society transforms biological sexuality into products of human activity."[12] That is, Kessler did not see the cultural construction of gender as arising out of a "sex/gender system" wherein "sex" functions as the precultural, biological base on which cultural (or superstructural) gender is overlaid. Rather, Kessler figured the concept of a foundational, essential, innate, true, and abiding "sex" as a backformation or retroactive effect of the cultural construction of gender. "Case management involves perpetuating the notion that good medical decisions are based on interpretations of the infant's real 'sex' rather than on cultural understandings of gender."[13] Yet ideas about what constitutes the appropriate shape, size, and function of "male" versus "female" genitals and anatomies are culturally determined. Kessler explains:

> In the face of apparently incontrovertible evidence—infants born with some combination of "female" and "male" reproductive and sexual features—physicians hold an incorrigible belief in and insistence upon female and male as the only "natural" options. This paradox highlights and calls into question the idea that female and male are biological givens compelling a culture of two genders.[14]

For Kessler, the medical treatment of intersex people is paradoxical because, on the one hand, it acknowledges that sexual dimorphism is not a given, while, on the other hand, it uses surgical and hormonal technologies to reshape non-dimorphic bodies to become dimorphic, literally engendering them. "The nonnormative is converted into the normative, and the normative state is considered natural."[15] Kessler concludes that, with the exception of rare cases where the health of the infant is genuinely threatened, in general "genital ambiguity is 'corrected' not because it is threatening to the infant's life but because it is threatening to the infant's culture."[16]

From this perspective, what intersexuality threatens is not only the widely held cultural belief that sex is naturally dimorphic, or that sexual dimorphism is mandated by biology, but, more radically perhaps, the assumption that dimorphic sex both precedes and is the origin of gender. What gets to count as an intelligible and legible "male" or "female" body, is, Kessler implies, constituted by cultural technologies rather than biology alone. By rethinking gender's cultural construction through gender's surgical reconstruction, Kessler seeks to link the ideological systems of meaning making that inform normative conceptions of gender with technological and biomedical practices that enable intervention into the materiality of bodies. Reading for ideology's material effects, Kessler's project not only stages an important intervention into feminist theories of sex/gender, but also counters the popular misconception that gender's cultural construction is either a purely discursive or sociological process. Gender is literally inscribed into the flesh of situated bodies.

In making this claim, Kessler is staging an argument with fellow feminist theorists over how to conceptualize the sex/gender distinction in relation to (what she saw as) the false dichotomy of nature versus culture. Citing Sherry Ortner, Kessler points out that "the nature/culture distinction is itself a construction—a product of culture."[17] For Kessler, the point is not that bodies do not matter, empirically or biologically speaking. Rather, the point is that biology is a discipline of knowledge produced by human activity, not a transcendent and transhistorical regime of truth.

Etymologically, "biology" combines *bios* with *logos*. As Foucault reminds us, biology becomes a modern science and field of knowledge by targeting life as the object of techniques.[18] Drawing on this important Foucauldian point, Kessler argues that the biomedical sciences produce knowledge of human bodies by interpreting and reading bodies through the cultural prism of gender normativity. In the case of intersex infants, Kessler notes,

> Physicians . . . consider several factors beside biological ones in determining, assigning, and announcing the gender of a particular infant. Indeed, biological factors are often preempted in their deliberations by such *cultural* factors as the "correct" length of the penis and capacity of the vagina.[19]

In naming "factors such as the 'correct' length of the penis and capacity of the vagina" as *cultural*, Kessler emphasizes that these are not naturally given qualities but rather heteronormative cultural assumptions, and these conventions materially regulate which bodies get to count as intelligible and which do not.

Though she emphasizes the regulatory role played by the biomedical sciences in shaping what gets to count as a legibly gendered body, Kessler also attends to the possibilities of agency and transformation that arise in relation to the medical and cultural regulation of gender.

> If authenticity for gender resides not in a discoverable nature but in someone's proclamation, then the power to proclaim something else is available. If physicians recognized that implicit in their management of gender is the notion that finally, and always, people construct gender as well as the social systems that are grounded in gender-based concepts, the possibilities for real societal transformations would be unlimited.[20]

Kessler goes on to suggest that the recognition of gender as culturally constructed has ethical consequences, specifically with regard to questions of accountability. In the conclusion to *Lessons of the Intersexed*, Kessler argues that the medical treatment of intersex folks reveals that gendering is a pervasive social practice that reproduces relations of power and historically contingent though profoundly real hierarchies of domination. "If intersexuality imparts any lesson," Kessler writes, "it is that gender is a responsibility and a burden—for those being categorized and for those doing the categorizing."[21] To be read as a gendered being, to attribute gender to others, is to participate in a material-semiotic network where speech acts, categorical claims, behaviors, and actions have ethical and political implications. The intersexed are simultaneously posited as "other" to dominant understandings of sex and gender, and surgically normalized in the name of recuperation and health. In her polemically charged conclusion, Kessler calls on members of society "to use whatever means we have to give up on gender,"[22] but she provides neither a concrete set of strategies for doing so nor a rationale for why this is the ultimate "lesson" she draws from the intersexed.

Fausto-Sterling and Developmental Systems Theory

Whereas Kessler theorizes "biological sex" as an effect of the cultural construction of gender, Anne Fausto-Sterling, another of the first feminist scholars to consider the causes and effects of the medical management of intersexuality, uses the very data of biology to call the "nature" of sexual dimorphism into question. In her 1993 essay, "The Five Sexes: Why Male and Female Are Not Enough," Fausto-Sterling observes that, "in the idealized, Platonic, biological world, human beings are divided into two kinds: a perfectly dimorphic species."[23] However, "on close inspection, absolute dimorphism disintegrates even at the level of basic biology."[24]

> If the state and the legal system have an interest in maintaining a two-party sexual system, they are in defiance of nature. For biologically speaking, there are many gradations running from female to male; and depending on how one calls the shots, one can argue that along that spectrum lie at least five sexes—and perhaps even more.[25]

Importantly, Kessler points out in *Lessons of the Intersexed* that Fausto-Sterling's 1993 five-sex model still gives precedence to genital morphology as the determining factor of sex.[26] In her 2000 return to the topic, "The Five Sexes, Revisited," Fausto-Sterling recognizes the validity of Kessler's critique and suggests that it would be better for intersex people and their supporters to "turn everyone's attention away from genitals," to acknowledge that people's anatomical morphologies are remarkably diverse, and that genitals alone cannot adequately define the meaning of either gender or sex.[27]

Whereas Kessler privileges culture above biology in her analysis of the medical treatment of intersexuality, Fausto-Sterling's work articulates a strong concern with understanding the complex interactions between biology and culture. If Kessler adopts what we might call a radical constructionist framework, Fausto-Sterling, by contrast, has sought to offer an approach she calls interactive, focusing on how biological and cultural processes are entangled and mutually constitutive.

In her 2000 monograph *Sexing the Body: Gender Politics and the Construction of Sexuality*, Fausto-Sterling combines feminist theory with the history of science to investigate the ways in which various scientific disciplines—endocrinology, genetics, neuroscience, and other fields—produce knowledge about gender, sex, and sexuality and argues for the necessity of moving beyond the dualisms of nature/culture, sex/gender, male/female, and heterosexuality/homosexuality.[28] Using a "developmental systems theory" approach, Fausto-Sterling asks how sex, gender, and sexuality are shaped by interrelated and overlapping cultural and biological systems. Fausto-Sterling suggests that no account of embodiment can afford to simply discount the role played by biological processes, yet she is careful to point out that biological processes are shaped by cultural contexts, institutions, and processes. For Fausto-Sterling, hormones, genes, chromosomes, and other biological features are real, material elements of human physiology. They are not simply products of ideology, but neither are the meanings they are given in any particular cultural and historical context free from ideological overdetermination. Drawing on her training as a biologist and pioneer of feminist science studies, Fausto-Sterling contends that the relation between biology and culture is not a simple opposition. As she puts it in the introduction to *Sexing the Body*,

> truths about human sexuality created by scholars in general and biologists in particular are one component of political, social, and moral struggles about our cultures and economies. At the same time, components of our political, social, and moral struggles become, quite literally, embodied, incorporated into our very physiological being. My intent is to show how these mutually dependent claims work, in part by addressing issues such as how—through their daily lives, experiments, and medical practices—scientists create truths about human sexuality; how our bodies incorporate and confirm these truths; and how these truths, sculpted by the social milieu in which biologists practice their trade, in turn refashion our cultural environment.[29]

By foregrounding the ongoing chain of interactions between biology and culture, Fausto-Sterling offers a strikingly original and dynamic alternative to the standard framework of "nature versus nurture" or "essentialism versus constructionism." Her work makes it possible to see that the sexing of the body is a process wherein the biological becomes cultural and the cultural, in turn, becomes biological.

Biology, in this account, is something more than just the raw "stuff" bodies are made of. For Fausto-Sterling, biology refers not only to the materiality of bodies, but also to a scientific set of discourses and practices embodied in particular human institutions. Though Fausto-Sterling seeks to provide a more nuanced account of the role of biology in the sexing of the body, she concurs with Kessler about the social basis of ideas about sex and gender. A close reading of the history of the biological sciences reveals that

> labeling someone a man or a woman is a social decision. We may use scientific knowledge to help us make the decision, but only our beliefs about gender—not science—can define our sex. Furthermore, our beliefs about gender affect what kinds of knowledge scientists produce about sex in the first place.[30]

According to this line of thought, sex, even at the biological level, does not have a pure, unmediated, unequivocal referent. It materializes through both the research process and the social and ideological contexts in which researchers live and breathe. "Beliefs about gender," Fausto-Sterling stresses in the above passage, are inseparable from the ways in which sex is defined. This does not imply that "sex" is either immaterial or unreal. On the contrary, it suggests that "sex" is a product of the material-semiotic power structures that compose and regulate biocultural formations. In this sense, Fausto-Sterling's argument implies that sex and gender cannot be fully or finally unraveled from one another.

> Our bodies are too complex to provide clear-cut answers about sexual difference. The more we look for a simple physical basis for "sex," the more it becomes clear that "sex" is not a purely physical category. What bodily signals and functions we define as male or female come already entangled in our ideas about gender.[31]

Suggesting that sex cannot be definitively disentangled from gender, Fausto-Sterling turns to the history of the medicalization of intersex people to show that research agendas are shaped by both explicit and implicit cultural and political agendas, which in turn shape the very ways in which bodies come to matter.[32]

One of Fausto-Sterling's most important contributions to the story of sex and gender through the lens of intersex is her attention to the role played by changing technologies and scientific practices in the shifts in our beliefs about sex and gender. Fausto-Sterling dedicates several chapters of *Sexing the Body* to the history and politics of biomedical research on intersexuality. She observes that it was not until the 1950s, which saw the development of surgical technologies capable of changing the body's morphology, that medical experts began to treat intersex as an anatomical pathology that could be surgically "corrected." Today, the dominant clinical view understands intersex conditions as "correctable" health "disorders," but Fausto-Sterling stresses that physicians could just as easily understand intersex states as a natural expression of human anatomical diversity. Training her attention on the power relations that structure scientific and medical discourses and practices, Fausto-Sterling argues that "intersexuals, seen as deviations from the norm who need to be 'fixed' in order to preserve a two-gender system, are also studied [by medical professionals] to prove how 'natural' the system is to begin with."[33] Contradictions such as this indicate that medical science does not merely reflect cultural norms but rather contributes to their production through its own supposedly "value-free" practices. Critically examining John Money's research on intersexuality, Fausto-Sterling emphasizes that sexual difference does not exist in a historical vacuum. Sexual difference is not a thing, she argues. Rather, it is a material-semiotic practice, one that human institutions—western biomedicine and its globalizing cultural circuits of influence, in particular—inscribe onto bodies.

> Until very recently, the specter of intersexuality has spurred us to police bodies of indeterminate sex. Rather than force us to admit the social nature of our ideas about sexual difference, our ever more sophisticated medical technology has allowed us, by its attempt to render such bodies male or female, to insist that people are either naturally male or female. Such insistence occurs even though intersexual births occur with remarkably high frequency and may be on the increase. The paradoxes inherent in such reasoning, however, continue to haunt mainstream medicine, surfacing over and over in both scholarly debates and grassroots activism around sexual identities.[34]

Affirming intersex activist critiques of nonconsensual infant genital surgeries and hormonal treatments, Fausto-Sterling contends that the medical management of intersex needs to change to protect not only patients' rights but also patients' long-term health.[35] She details the specific practices that activists are pursuing to institute those changes. And finally, Fausto-Sterling notes that a growing number of physicians have already begun to accept the criticisms of intersex adults and activists and have started to rethink the praxis of intersex medicine. These developments, Fausto-Sterling concludes in a somewhat utopian vein, suggest that "we are moving from an era of sexual dimorphism to one of variety beyond the number two."[36]

Butler and Gender Performativity

In suggesting that contemporary social formations are already moving toward a more fluid and diverse understanding of gender, Fausto-Sterling implies that the proliferation of gender contestations—of occasions when dimorphic gender's meaning and materiality are thrown into radical question—holds the potential to profoundly reshape social relations. This point is at the heart of Judith Butler's highly influential 1990 monograph *Gender Trouble: Feminism and the Subversion of Identity*. Butler asks how the regulatory operations of what she calls, after Adrienne Rich, "compulsory heterosexuality" maintain various forms of sexual hierarchy. In so doing, she stages an important intervention into feminist debates over the sex/gender distinction by arguing against the notion that biological sex is the foundation of cultural gender, that sex forms the "natural" substance onto which the social meaning of gender is written. Rather, "gender ought not to be conceived merely as the cultural inscription of meaning on a pregiven sex (a juridical conception); gender must also designate the very apparatus of production whereby the sexes themselves are established."[37] Butler continues in a sentence that has become famous: "As a result, gender is not to culture as sex is to nature; gender is also the discursive-cultural means by which 'sexed nature' or 'a natural sex' is produced and established as 'prediscursive,' prior to culture, a politically neutral surface *on which* culture acts."[38] In these formulations, Butler suggests that gender constitutes a social apparatus that naturalizes the illusion of a prediscursive sex. "Sex itself is a gendered category," Butler writes, thereby defining "sex" as an effect rather than the cause or ground of gender.[39]

In arguing that discursive gender produces the notion of prediscursive sex and that gender is performative, Butler articulates a powerful deconstructive reading of the nature/culture opposition that differs in key respects from both Kessler's radical constructionist and Fausto-Sterling's interactionist approaches. As Diane Elam summarizes Butler's position, "nature [for Butler] is the retro-projected illusion of a real origin to culture, yet that illusion is *necessary* to culture, the very ground of its capacity to represent itself."[40] The deconstruction of the nature/culture opposition allows Butler to conclude that "the very notion of an essential sex and a true or abiding masculinity or femininity are also constituted as part of the strategy that conceals gender's performative character and the performative possibilities for proliferating gender configurations outside the restricting frames of masculinist domination and compulsory heterosexuality."[41]

Gender Trouble is important not only for its demystification of sex/gender but also for its attention to intersex specifically. As Robyn Wiegman notes,[42] though most critical commentary on *Gender Trouble* has focused on Butler's usage of the figure of drag to explain the theory of performativity, Butler's account of gender performativity actually dedicates far more space to a consideration of the question of hermaphroditism in *Herculine Barbin, Being the Recently Discovered Memoirs of a Nineteenth Century Hermaphrodite*.[43] Butler encountered Barbin through the work of Michel Foucault, who published Barbin's autobiography along with an editorial "Introduction" in English in 1980. According to Wiegman, "Butler's discussion of Barbin . . . was crucial to *Gender Trouble*'s analysis of the heterosexual regulatory strategies that normatively align sex, gender, and sexuality."[44] Wiegman explains: "The intersexuality of Barbin's genitals and her/his passage in the course of a truncated lifetime through the categorical designations of both male and female challenged any definitive account of the trajectory of sexual desire—was she/he homosexual or heterosexual, seemingly 'normal' or sexually 'deviant'?"[45] Butler uses Barbin's narrative to provide a vivid illustration of the legal, medical, and psychosocial strategies of regulation that arise in relation to occasions of gender trouble. Further, she contends that Barbin's autobiography reveals that gender should not be conceived as a substantive identity, but rather as a process, a kind of ongoing doing, what Butler calls "a constituted social *temporality*."[46] Through Barbin's narrative about life in a nineteenth-century French convent and the legal and medical struggles that took place over her identity once it was discovered that she was not a "true" woman, Butler is able to argue that gender is "performative" in the sense that it is tenuously constituted by the very acts that are said merely to express it.

Importantly, Butler frames her argument as a repudiation of what she sees as the key claim of Foucault's editorial "Introduction" to the English-language edition of Barbin's autobiography: that hermaphroditism represents the "happy limbo of a non-identity,"[47] a claim that implies, according to Wiegman's reading of Butler, that Barbin existed "outside the law precisely because of the body's failure to conform to the law's regulatory schema of dimorphic sex."[48]

Against this position, Butler contends that the legal regulation of Barbin's narrative demonstrates precisely the performativity of gender. By theorizing hermaphroditism as central to questions of gender regulation, Butler suggests that normative conceptions of gender are tied to social and political processes that exclude particular forms of embodiment and personhood from the domain of the intelligibly human.

While it is clear that intersexuality plays a key role in Butler's account of gender performativity, it seems worth pausing here to contemplate whether Butler's reading of Foucault and Wiegman's reading of Butler accurately reflect Foucault's own position as staked out in his "Introduction" to Barbin's memoir. Interestingly, Foucault's "Introduction" to the volume was only published in the English edition of the book, translated by Richard McDougall, which appeared in 1980. The French edition, published in 1978, only contains Barbin's text itself.[49] What became the "Introduction" to the English edition was originally published in French in a slightly but crucially different form in the journal *Arcadie* in 1980.[50] Considering Foucault's long-standing interest in theorizing the complexities of power relations, and considering his lectures at the Collège de France from 1974–75, *Abnormal*, published in English for the first time in 2004, where Foucault provides a detailed study of historical shifts in the medical regulation and juridical treatment of hermaphroditism, it never quite made sense to me that Foucault would frame hermaphroditism as existing outside of power relations, as Butler claims, as if Foucault had momentarily forgotten his own theorizations of power.

Specifically, Butler argues that the journals of Barbin and their "Introduction,"

> offer an occasion to consider Foucault's reading of Herculine against his theory of sexuality in *The History of Sexuality, Volume 1*. Although he argues in *The History of Sexuality* that sexuality is coextensive with power, he fails to recognize the concrete relations of power that both construct and condemn Herculine's sexuality. Indeed, he appears to romanticize h/er world as the "happy limbo of a non-identity,"[51] a world that exceeds the categories of sex and identity.[52]

The line Butler highlights is framed within the following paragraph from Foucault's English "Introduction," which I quote in full.

> Alexina wrote her memoirs about [her former life] once her new identity had been discovered and established. Her "true"

> and "definitive" identity. But it is clear that she did not write them from the point of view of that sex which had at least been brought to light. It is not a man who is speaking, trying to recall his sensations and his life as they were at the time when he was not yet "himself." When Alexina composed her memoirs, she was not far from her suicide; for herself, she was still without a definite sex, but she was deprived of the delights she experienced in not having one, or in not entirely having the same sex as the girls among whom she lived and whom she loved and desired so much. And what she evokes in her past is the happy limbo of a non-identity, which was paradoxically protected by the life of those closed, narrow, and intimate societies where one has the strange happiness, which is at the same time obligatory and forbidden, of being acquainted with only one sex.[53]

The line "the happy limbo of non-identity" appears in French as "les limbes heureuses d'une non-identité."[54] The irony of a "happy limbo" is evident enough in the English but has an even stronger resonance in the French. *Limbo/les/limbes* has a theological connotation (especially in French): etymologically, "limbo" means a state wherein one is at the edge of hell but locked out of paradise. It is a state of existence that is heavily constrained. From this perspective, it is hard to see how this state of being "locked up" in-between could be purely "happy." Indeed, the "locked up" state of a "happy limbo" can be read as a mirror of the "strange happiness" of the "closed, narrow" world of the convent in which Barbin lived and worked as "Alexina" until her hermaphroditism was discovered and she was obliged to make a legal change of sex to being a "man" after juridical proceedings. Foucault's characterization of Barbin's memoir as evoking "the happy limbo of a non-identity" thus contains within it an ironic acknowledgement of the circumscribed subject-position Barbin occupied in two spaces: both within the constrained, same-sex convent and outside it, Barbin's life is ruled by the law of sexual dimorphism. In the former, it meant a world of "monosexuality" (a world only for women), whereas in the latter it meant a strict heterosexual binarism. In either case, Barbin's text hardly represents the bucolic, romantic vision Butler paints.

More interestingly, in the French original there is a long paragraph, completely omitted from the English translation, about the use of "discretion" by directors of conscience in religious institutions. Foucault describes "discretion" as having a double meaning: on the one hand, it means the capacity to perceive differences, to ferret out feelings and the impurity of souls, to separate the passion that comes from God and that which is inculcated by the Seducer (i.e., the Devil). At the same time, "discretion" also means for these directors of conscience the ability to maintain a certain measure, to restrain oneself, to not go too far, to keep unspoken that which should not be spoken, to leave in the shadows that which would be dangerous in the light of day.

> One might say that Alexina was able to live for a long time in this chiaroscuro ["clair obscur," which literally translates as "light dark"] of a regime of "discretion" which was that of convents, boarding schools, and feminine Christian monosexuality. And then—this was her drama—she passed under another regime of "discretion." That of administration, justice, and medicine. The nuances, the subtle differences that were recognized in the first no longer held. But that which was kept silent in the first had to be clearly shared in the second. To tell the truth, this is no longer about discretion but analysis.[55]

In moving from one regime of "discretion" to another, Alexina moved from the rules of the convent to a world organized by "administration, justice, and medicine." This shift was both disciplinary and biopolitical. It was disciplinary in the sense that Barbin's individual identity and body became an object of explicit state and medical regulation. And it was biopolitical in that, by assigning Barbin a "true" and "definite" identity as a man, the magistrates maintained sexual dimorphism as the law of populations, even or perhaps particularly in "monosexual" spaces such as the convent. Thus Barbin, as a person assigned to the category of maleness, could no longer belong to the world of the convent. From this perspective, Foucault's claim that Barbin's narrative traces the "happy limbo of non-identity" does not ascribe to Barbin a status outside gender, sex, the law, and power relations, but rather frames Barbin's subject position *within the convent* as regulated precisely in terms of Barbin's status as an "other" of sexual difference, a subject position both produced and ultimately foreclosed by the cultural logic of sexual dimorphism.

My critique of the nuances of Butler's analysis notwithstanding, my reading of Foucault's "Introduction" seeks to build on what I believe is ultimately Butler's larger point, that "concrete relations of power . . . both construct and condemn Herculine's sexuality."[56] In particular, it was the uncertainties Barbin's body provoked—was she/he female or male, and, as Wiegman asks, "homosexual or heterosexual, seemingly 'normal' or sexually 'deviant' "—that both medicine and the law refused to tolerate.[57]

The Barbin controversy reveals the political stakes of the destabilization of sex, gender, and sexuality binaries and thereby opens the question of what forms of sociality would enable us to preserve, rather than legally and medically abolish, those uncertainties.

Against Intersex Exceptionalism

The three previous sections of this chapter sought to demonstrate the specific ways in which Kessler, Fausto-Sterling, and Butler analyzed intersexuality to rethink feminist and queer analytics of sex, gender, and sexuality. In this section, I shift to consider some of the problems generated by their foundational interventions. In their essay "One Percent on the Burn Chart: Gender, Genitals, and Hermaphrodites with Attitude," David Valentine and Riki Anne Wilchins argue that the ethicopolitical implications of the uses of intersex in feminist and queer scholarship have not received enough critical attention.[58] Valentine and Wilchins call on feminist and queer scholars to move "beyond thinking about intersex and transgender bodies as some kind of literal performative that neatly shows how gender and the body are discursively produced." They suggest instead that this position should be "a starting point for thinking about what all bodies mean and can mean in different contexts, and how this meaning is enforced," not the end point of analysis.[59] "Bodies which are suspect," they caution, "are not what have to be explained. Rather, the requirement that they explain themselves should itself be investigated."[60]

The early feminist literature on intersexuality as exemplified by Kessler, Fausto-Sterling, and Butler can be understood, in part, as a critique of the notion that intersex bodies require explanation. But because these theorists are invested in *explaining* why intersex bodies have disproportionately been subject to the techniques of biomedical regulation, their work (and the work of scholars who follow in their footsteps, a group in which I implicate myself) has also tacitly participated in the production of a kind of intersex exceptionalism. Intersex exceptionalism is the view that intersex bodies are historically extraordinary, isometric objects of study—objects like no others—that reveal spectacular truths. Intersex exceptionalism is problematic because it presumes that "minus a few exceptions the system works just fine."[61] That is to say, intersex exceptionalism hypostatizes ideas about the nature of atypical sex and gender nonconformity and, on the other hand, renaturalizes—rather than calls into question—the embodiments of people with non-intersex anatomies and cisgender or gender-conforming presentations, as well as the gendered operations of medical, legal, and social systems more broadly.

The problem of intersex exceptionalism is evident in different ways in Kessler, Fausto-Sterling, and Butler. While each critiques the medical normalization of intersex folks and challenges medicine's appropriation of intersex bodies to ideologically naturalize sexual dimorphism, they concurrently seem to suggest that the history of intersex is exceptional in comparison to more familiar histories of the biopolitics of sex and gender. This exceptionalism substantiates, in Kessler's case, a radical constructionist framework; in Fausto-Sterling's, an interactionist model of biocultural systems development; and in Butler's, the theory of gender performativity. Kessler, Fausto-Sterling, and Butler each treats intersex as an exceptional kind of evidence. They then interpret that evidence to make claims about how intersex *singularly* demonstrates the instability and complexity of gender binarism and sexual dimorphism, almost as if other (normative and non-normative) configurations of embodiment do not also reveal that instability and complexity. In this respect, their arguments invert but do not displace the hegemonic biomedical paradigm John Money inaugurated. Reversing Money's hierarchy, Kessler, Fausto-Sterling, and Butler each holds intersex bodies up as emblems of human variation and suggests that progressive thinkers and activists ought to embrace, even celebrate, such variation in order to expand the range of possibilities for what gets to count as an intelligible and livable human life.

This suggestion's political optimism situates sex variation and gender diversity as positive sociopolitical ideals aligned with multiculturalism, LGBT acceptance and visibility, immigrant rights, and other liberal movements that seek inclusion within the nation-state. However, there are significant risks implicit but generally unacknowledged in the elevation of sexual variation and gender diversity into such an ideal. As scholars of women of color feminisms and queer of color critique argue, liberal subjectivity secures freedom for socially and economically privileged populations while, at the same time, disproportionately targeting marginalized populations—especially women (cis- and trans) and queer people of color, as well as undocumented and poor folks—with the daily threat and recurrent practice of various forms of structural violence.[62] For this reason, it is important to ask what roles sex/gender diversity has come to play in neoliberal regimes of security and surveillance and international divisions of labor.

In addition, the intersex exceptionalism apparent in Kessler, Fausto-Sterling, and Butler is also problematic because it can reify ideas about the nature of intersex alterity. In her monograph *Intersex: A Perilous Difference*,

Morgan Holmes provides an extended analysis of this conundrum.[63] According to Holmes, there is a tendency in some feminist and queer work to fetishize intersexual difference as a form of radical otherness. Holmes argues that this fetishization presumes that intersex folks somehow have a more "natural" responsibility to be critical of sexual and gendered norms than other folks do. In her reading of Sharon E. Preves's ethnographic sociology of the medicalization of intersex and intersex political activism,[64] Holmes takes issue with the way in which Preves valorizes "the intersexed subject as a kind of gender-savant."[65] According to Holmes, "it appears that Preves expects her intersexed participants to lead western culture out of the darkness of a two-sex system."[66] Preves's argument, Holmes suggests, "creates an ideal intersexed subject whose perceptions of embodiment and ability to detect gender dogma will lead the way to enlightenment."[67] This position, Holmes observes,

> lacks compassion for those who do not maintain a critical relationship to the operation of gender norms or of heteronormativity. For this reader it feels as though having been identified as a statistical outlier I and others like me, it seems to say, cannot be permitted to want simply to be like all the other girls and boys; we must instead lead the charge against taking gender for granted.[68]

In assuming that the category of intersex experience is transparent and homogeneous, and an innately radical challenge to the binary sex/gender system, Preves attributes to intersex people an obligation "to willingly and gladly inhabit a space of resistant unintelligibility."[69] This attribution not only denies the agency of intersex people to articulate their own subject positions. It also frames those intersex folks who seek to live "normal" lives—for instance, by organizing their existence around culturally white, middle-class, and heteronormative forms of gendered and sexual expression (such as the nuclear family, monogamy, consumer culture, and so forth)—as politically regressive and insufficiently radical. According to Holmes, Preves presumes that power is a zero-sum game. This leads her to make moralizing claims about intersex subjects who take part in normative behaviors and structures without providing an account of the complicated ways in which those subjects make negotiations and enact forms of agency in their daily lives that cannot be captured by an overarching binary of liberation versus subordination. In so doing, Holmes argues, Preves imposes a normative/non-normative binary as the exclusive means by which to understand the sociopolitical field, an imposition that forecloses more nuanced accounts of normativity's complexity and ultimately upholds the very logic of normalization her project seeks to call into question.

Holmes's analysis is useful because it gives pause to the tendency to read intersex embodiment as a simple synonym for queer gender. Holmes points out that "a critical relationship to the operation of gender norms or of heteronormativity" is not a necessary or essential feature of any particular form of embodiment. Rather, such a relationship, where developed, is possible because it is an outcome of particular processes of subjectification and subjection. "It is critical to remember," Holmes writes, "that whatever the perception of the individual regarding his/her condition and treatment, none is *obliged* to act as an advocate for non-normative agendas."[70] For Holmes, intersex, like any sexed or gendered category, does not *necessarily* prefigure any predetermined political consciousness or standpoint. Rather, such a standpoint, where developed, is the product of specific forms of epistemic and political struggle. Figuring intersex embodiment as intrinsically subversive treats intersex people as objects but not subjects of analysis and politics. Paradoxically, this view ends up replicating the very structure of the subject/object division on which the medicalization of intersexuality is based.

Taken together, the problems of intersex exceptionalism and the fetishization of intersex alterity direct our attention to the ethics of producing knowledge about intersex.[71] In this regard, perspectives that emphasize self-reflexivity, critical positionality, and social location can help to temper these problems by refiguring critical intersex studies as part of a larger web of allied intellectual projects that seek to understand the complexities of power that shape diverse people's lives.

Conclusion

The scholarship I have analyzed in this chapter suggests in different ways that critical intersex studies productively troubles long-standing feminist paradigms. While some might interpret this trouble as the epistemological undoing of women's studies and the various field formations it inaugurated,[72] I take another view. The troubling of sex/gender not only contests hegemonic and essentialist iterations of sexual difference, but it also contributes to the broader effort to critically reconsider women's and gender studies' complicated and fraught relationships with, and variegated investments in, sexual dimorphism as ground of being, knowing, desire, agency, and politics. This reconsideration creates opportunities to move toward what trans theorist Jean Bobby Noble calls, after Jordy Jones, "genders without genitals"—that is, a conception of gendered embodiment detached from the presumption of dimorphic genital referentiality.[73] Furthermore, by critically interrogating the sex/gender and biology/culture binaries and interrupting the equation of genitals with sex and gender, critical intersex studies complexifies our understandings of sexual dimorphism, gender binarism, and heteronormativity, opening new possibilities for theorizing sex, gender, and sexuality as intersectionally constituted performative systems of corporeal inscription.

In challenging the assumed naturalness, stability, and coherence of sexual dimorphism, critical intersex studies certainly issues a profound challenge to entrenched certainties about embodiment. At the same time, it also suggests that uncertainty about the objects/referents of feminist and queer research is ethical. Such uncertainty is ethical because it maintains openness to the unknown and the unpredictable, to possibilities, materialities, and temporalities that remain unanalyzed or yet to come. Thus, rather than precluding the study or usage of the word *women* in women's, gender, and sexuality studies, as the search committee member from the anecdote I began this chapter with worried, the critical study of intersex complexifies the meaning of that word, calling attention to the exclusionary and regulatory effects of the material-semiotic reproduction of sexual dimorphism, gender binarism, and heteronormativity.

My argument here takes its inspiration, in part, from Gayle Salamon's essay "Transfeminism and the Future of Gender."[74] In her reading of the relation between contemporary women's studies and the developing and newly institutionalizing field of trans studies, Salamon suggests that transgender phenomena can be read as a challenge to the definitional stability of women's studies as a disciplinary project. Explaining the resistance to transgender studies among some feminist academics, Salamon writes, "the category of 'woman,' even if it is understood to be intersectional and historically contingent, must offer a certain persistence and coherence if it is to be not only the object of study but the foundation of a discipline, and a subject formation that describes a position of referential resistance might not be easily incorporated into such a schema."[75] Salamon's theorization suggests that some though not all trans phenomenologies resist the referential grid of binary gender. But rather than reading these elements of trans studies as necessarily opposed to women's studies, Salamon argues that the intellectual domain of women's studies, if it is to remain critical, must remain open to resignification, and that trans studies has much to contribute to the project of rethinking and rearticulating the analytic, epistemological, and political horizons of women's, gender, and sexuality studies. Likewise, Salamon also argues that trans studies can benefit greatly from the rich history of feminist analyses of gender through intersectional and transnational lenses.

Following Salamon's lead, I understand critical intersex studies as both an ally to and a critical interlocutor for feminist, queer, and trans studies alike. In particular, intersex studies reframes the western medical pathologization of atypical sex as a condition of the materializations of all manifestations of gender in western/global northern societies and their globalizing channels of power, even and especially including cisnormative, heteronormative, homonormative, and transnormative narratives of being that obscure the centrality of the medicalization of intersex to the genealogy of gender as a paradigm of subjectivity and sociality in the modern era. This chapter has therefore sought to extend the argument I made in chapter 1 by emphasizing that feminist scholarship still has much to learn from the various "sexed others" against which and in relation to hegemonic systems of bodily categorization take shape. Learning from the history of intersex is a process of unlearning received accounts of gender's history. As Morland puts it in the epigraph with which I opened this chapter, this unlearning might be used to support an ethics that preserves the uncertainties intersex bodies provoke.

3

"Stigma and Trauma, Not Gender"

A Genealogy of US Intersex Activism

> Since when are intersexuals necessarily interested in subverting anything?
>
> —Morgan Holmes[1]

In her groundbreaking 1998 article "Hermaphrodites with Attitude: Mapping the Emergence of Intersex Political Activism," Cheryl Chase—the founder and longtime executive director of the Intersex Society of North America (ISNA)—sketched a detailed chronicle of the emergence of intersex political activism in North America in the early 1990s, focusing on the struggle to garner a broad base of support for destigmatizing intersexuality and for reforming the dominant medical model of intersex management.[2] While the majority of health care professionals were wary of and sometimes outwardly hostile to the claims of the burgeoning intersex movement, Chase singled out "some gender theory scholars, feminist critics of science, medical historians, and anthropologists" as having been uniquely receptive to and "quick to understand and support intersex activism."[3] Chase cited, among others, Suzanne J. Kessler and Anne Fausto-Sterling. As I noted in earlier chapters, Kessler was one of the first scholars to suggest, in her 1990 *Signs* article, that gendered norms and expectations shape the medical treatment of intersex infants, while Fausto-Sterling's 1993 *Sciences* essay "The Five Sexes" took issue with the presumption of sexual dimorphism in western science and culture, thereby opening the door for Chase's inaugural public announcement of the formation of ISNA in a response letter published in the next issue of *Sciences*.[4] Whereas more mainstream feminist organizations such as NOW and the Feminist Majority Foundation were slower to endorse the intersex movement,[5] Kessler, Fausto-Sterling, and a select few other scholars including Alice Dreger were, in Chase's words, "early ISNA allies."[6]

As I argued in the last chapter, these feminists questioned the assumptions underlying the dominant biomedical approach to intersex treatment. They suggested that the medical normalization of people diagnosed as intersex was part of a broader regulatory regime of power-knowledge that sought to manage and contain the material-semiotic destabilizations of dimorphic sex. According to Chase, the work of Kessler, Fausto-Sterling, Dreger, and others provided crucial intellectual and political support for the intersex movement during its initial years, helping intersex activists to "command a measure of social legitimacy" during a time when there was tremendous resistance to any effort to critically rethink the medicalization of intersex.[7]

Fast-forward just shy of a decade after "Hermaphrodites with Attitude" was published, and Chase's view of the relationship between feminist scholarship and the intersex movement changes dramatically. In a 2006 interview, Chase critiques "people in queer theory and women's studies" for imposing on intersex people the idea that their "identity should be determined by their anatomy" and for using intersex as an object of evidence to justify feminist and queer political desires for a radically different and differently gendered future.[8] In their 2002 essay published in *Women's Studies Quarterly* titled "From Social Construction to Social Justice: Transforming How We Teach about Intersexuality," Emi Koyama and Lisa Weasel provide a background to Chase's assessment.[9] Questioning what they call the appropriation of intersex for gender theory, Koyama and Weasel argue that "the political and practical issues relating to intersex lives have been marginalized in feminist scholars' use of intersex existence in support of their theoretical and pedagogical deconstructions."[10] As I discuss below, recent works by Alice Dreger and April Herndon, Vernon Rosario, and Katrina Karkazis echo and extend Chase's and Koyama and Weasel's critiques. In a paradoxical turn of events, a growing number of scholars and activists associated with the intersex movement have come to figure feminist scholarship—formerly understood as an *ally* to ISNA—as an obstacle to the intersex movement's political progress.

These charges expose a potentially significant rift emerging between intersex rights advocates and, on the other hand, practitioners of feminist and queer theory and women's, gender, and sexuality studies (WGSS). In an effort to rethink this divide, complicate the terms of the debate, and formulate alternatives, this chapter traces a genealogy of US intersex activism and critically analyzes the problems and possibilities of intersex critiques of women's studies, feminist theory, and queer theory.

Specifically, this chapter investigates the routinization of a certain gesture[11] within these intersex critiques: they routinely reiterate a claim made by Chase and ISNA that it is "stigma and trauma, not gender" that are the central factors that shape the problems intersex people face.[12] The above critics deploy this claim to construct feminist and queer scholarship on intersex as overly invested in theorizing and deconstructing gender at the expense of the practical and political issues related to intersex people's lived experiences. In so doing, I am concerned that these critiques may reify a problematic binary between theory and practice; and that they may also efface a richer and more complex genealogy of queer feminist engagements with intersex that not only challenge the medical pathologization of anatomical variation, but also reveal the role of gender regulation in shaping the interconnected systems of power that structure intersexed and non-intersexed people's lives.

At the outset, it is important for me to emphasize that I have learned a great deal from and am indebted to Chase, Koyama and Weasel, Dreger and Herndon, Rosario, and Karkazis. My own work would not be possible without their contributions. Moreover, I absolutely agree that if feminist research on intersex is to have a meaningful and lasting impact, it must be self-reflexive, attentive to local and global contexts, and informed by critical understandings of the lived experiences and material realities of intersex people. By identifying a routine but underinterrogated gesture within these authors' writings, I do not mean to reduce their contributions to the gesture that I am identifying. Furthermore, I do not want to give the false impression that their critiques share a single, common object or referent. I am not sure that they do. However, despite their different objects of critique, which they variously name "women's studies," "women's studies and queer theory," "feminist scholarship," and "gender theory" (all of which mean different things in different contexts but nonetheless remain linked to and are frequently used as synonyms for the field formation now known as WGSS), my analysis suggests that these writers consistently obscure the centrality of processes of gender regulation to intersex matters. In my view, the routinization of this gesture may foreclose opportunities to craft more nuanced genealogies of intersex and to form alliances across differences.

Through a variety of close readings, I argue, on the one hand, that these critiques tend to oversimplify feminist and queer work on intersex and obscure the genealogical linkages between the medical management of intersex and the biopolitical and geopolitical regulation of gender, sex, and sexuality in the twentieth and twenty-first centuries. On the other hand, it is my contention that these writers usefully expose a negativity or aggression that is difficult to bear because it is internal to feminist, queer, and intersex politics alike. Concurrently, I suggest that the intersex movement, in its many forms, can become more effective by taking seriously what Gayle Salamon calls "the systematic understanding that women's studies provides of the structures of gender—and the relations of power that underlie those structures."[13]

Chase's Shifting Views on "Women's Studies and Queer Theory"

As highlighted above, Chase's understanding of the relation between feminist scholarship and the intersex movement has changed considerably over the years. It is therefore worth examining the different views she has espoused and situating them in the context of her work as a founder of the intersex movement. Before doing so, it is important to underscore that there are many types of intersex activism. The intersex movement is multidimensional and complex. Though the movement originated in North America in the early 1990s, over the last two decades intersex activist groups have been founded in numerous countries around the globe.[14] What defines intersex activism is not necessarily a common politics (intersex activists hold diverse political views), nor a common identity (many do not frame intersex in identitarian terms); nor is it, as Karkazis points out, a common etiology (as "atypically" sexed individuals, just like "typically" sexed individuals, vary greatly in terms of physiological and biochemical traits and characteristics).[15] Instead, "what underpins the collective action of intersex activists . . . is [an awareness of] shared familial, social, and medical treatment experiences and subjective responses to these experiences—what medical anthropologists have broadly called the 'illness experience.' "[16] Since approximately the early twentieth century, intersexuality has been understood by western biomedicine as a set of discrete pathological conditions that results in various forms of deviation from the "normal" course of sex development.[17] Intersex activists have challenged this view by reflecting critically on the shared experience of the medicalization of individuals with atypically sexed bodies. However, as in any community, there is much disagreement among intersex activists, even over such seemingly basic matters as the definition of intersex, as well as questions of theory and practice, movement and coalition building, and long- and short-term movement goals. Considering the sheer heterogeneity of intersex politics, any account of the movement's history will be necessarily partial, open to contestation and revision.

Chase founded the Intersex Society of North America in San Francisco in 1993. ISNA started out as a grassroots support group but soon blossomed into a nonprofit organization whose mission was to transform the "standards of care" applied in the medical management of intersex infants, youth, and adults. Rooted in and influenced by "the queer politics of the 1990s," the intersex movement emerged as a politicized, oppositional stance against pathologization of non-normative bodies.[18] Just as early 1990s activists working on issues of sexuality and gender reclaimed the slur *queer* as a defiant critique of homophobic and heteronormative language, ideologies, institutions, and practices, ISNA likewise began by reappropriating the medical terms *hermaphroditism* and *intersex* and resignifying them.[19] Making intersex mean and matter differently than it had in the past was one component of ISNA's larger project of denaturalizing and disputing the dominant treatment protocols applied in the medical management of intersex. These protocols assert that infants born with intersex "conditions"—defined in terms of specific chromosomal, hormonal, gonadal, genital, and/or internal or external morphogenic anomalies—should be treated with genital normalization surgeries and/or hormone treatments so that their external morphology conforms with their binary sex/gender assignment.[20] Many intersex activists have argued and continue to argue that such treatments are, in the vast majority of cases, medically unnecessary and profoundly physically and psychologically harmful, stigmatizing, and traumatizing.[21] From 1993 until its disbandment in July 2008, ISNA spearheaded a multipronged campaign that included research, media outreach, lobbying efforts, and direct engagement with medical providers to call for the cessation of medically unnecessary infant genital surgeries. As the millennium drew closer, ISNA's work helped to spawn a broader intersex human rights movement around the globe. I trace a transnational genealogy of that movement in the next chapter.

In the early years of ISNA, Susan Stryker suggests,

> Chase considered intersex politics to be related to queer and transgender politics not only because they all challenged medical authority and called for the reform of powerful social institutions, but also because the practice of normalizing surgery was such a visceral example of the idea that beliefs about gender actually produce the sex of the body, rather than the other way around. Bodies that did not originally fit the gender binary were literally cut to fit into it.[22]

Stryker's comment accurately characterizes Chase's analysis of the medical treatment of infants and persons with intersex embodiments during the initial years of ISNA. Rather than considering intersex issues in isolation or in terms of a single-issue agenda, Chase recognized connections and points of overlap between queer, transgender, feminist, and intersex politics.

Moreover, Chase used these connections to frame an agenda for intersex activism that capaciously questioned not only the implications and so-called necessity of pediatric genital surgeries, but also the broader medical and social enforcement of normative beliefs about gender, sex, and sexuality.

To further grasp Chase's understanding of intersex politics in the early years of ISNA, I want to return to her widely read 1998 essay "Hermaphrodites with Attitude" to examine it in greater detail. "Hermaphrodites with Attitude" provides perhaps the clearest illustration of the way in which Chase used insights from feminist, queer, and transgender theory and politics to rethink the medicalization of intersexuality. In that essay, whose title Chase borrowed from the occasional newsletter published by ISNA from December 1994 through spring 2003, Chase begins by asking why physicians treat the birth of an infant with an intersex condition as a psychosocial emergency, one that supposedly requires swift surgical intervention. Chase writes: "Pediatric genital surgeries literalize what might otherwise be considered a theoretical operation: the attempted production of normatively sexed bodies and gendered subjects through constitutive acts of violence."[23] Chase connects material, institutionalized practices of intersex treatment to "what otherwise might be considered a theoretical operation." That is, she reads what actually happens in the clinic and the operating room as fundamentally connected to a specific theoretical operation. In naming pediatric genital surgeries "constitutive acts of violence," Chase further stakes a bold claim about intersex treatment, framing the dominant medical model of intersex management as a direct contravention of standards of medical ethics and the Hippocratic oath. Chase suggests that physicians are less concerned with the overall health and well-being of intersex infants and adolescents and more invested in literalizing dominant ideas about sex and gender by materially inscribing those ideas into the flesh of intersex bodies.

After providing a history of her involvement with the founding of ISNA, Chase goes on to offer an analysis of what she sees as the significant achievements of intersex activism in the 1990s. In its first years, ISNA used various political tactics to raise consciousness about the medical management of intersex. For instance, in 1996, ISNA members picketed the American Academy of Pediatrics Meeting in Boston. The protest was, to say the least, not well received by the medical community, and, after much consideration, ISNA decided to adopt a less "in your face" strategy. It was at this moment when ISNA began to formulate a patient-centered activist platform geared toward achieving practical medical reforms, a platform that centered on ending unnecessary, nonconsensual genital surgeries and improving clinical care for intersex children and adults. ISNA's medical reform agenda was designed as an effort to speak to medical providers in a language they could understand. By tailoring its activism to particular constituencies, ISNA's approach to social change can be understood as contextual and strategic. According to Chase, however, ISNA's medical reform strategy coexisted alongside what we might think of as the more *queer* elements of intersex activism. In this vein, Chase suggests in "Hermaphrodites with Attitude" that ISNA promoted intersex activism of many types, including the efforts of some more radical intersex activists to transform "intensely personal experiences of violation into collective opposition to the medical regulation of bodies that queer the foundations of heteronormative identifications and desires."[24]

As this last quotation from "Hermaphrodites with Attitude" indicates, Chase not only understood intersex activism as related to queer politics, but also adopted the language of queer theory to figure the existence of intersex people as a radical challenge to heteropatriarchal worldviews and structures. Emphasizing that the birth of an infant with a nonstandard anatomy calls into question naturalized expectations and assumptions about the meaning and materiality of human bodies, Chase suggests that intersex embodiments call the nature of sexual dimorphism into question, challenging widespread presuppositions about the natural occurrence, binarism, complementarity, ordering, and number of the sexes. Chase further hypothesizes that intersexuals "embody viscerally the truth of Judith Butler's dictum that 'sex,' the concept that accomplishes the materialization and naturalization of power-laden, culturally constructed differences, has really been 'gender all along.' "[25]

In laying claim to this Butlerian dictum, Chase reads the medical treatment of people with intersex anatomies as an exemplary case of the performativity of gendered embodiment. Pediatric genital surgeries do not merely or simply normalize the genitals of the infant. They cosmetically normalize the genitals of the infant while also often desensitizing them in an effort to create the likeness of a supposed "natural" dimorphic sex, which is then taken as the basis for a stable binary gender identity. So when physicians surgically correct the so-called nonstandard genitals of an intersex infant into aesthetically "normal" looking male or female genitals, they reconstruct the infant's sex in a way that will fundamentally shape the infant's life in myriad ways. As Iain Morland observes, surgery frequently entails the partial or total loss of capacity for genital sensation, a loss of self that is also a loss of others.[26] So even as surgical normalization literally attempts to perform gender on the infant, it also regularly fails in that performance, which is to say that it forecloses many intersex individuals' physical facility to experience sexual and other types of feeling. For Chase, the medicalization of intersex people must be understood as a specifically *violent* gendered phenomenon. In other words, Chase does not argue that intersex itself is a source of gender trouble, but rather that the medicalization of intersex figures the intersex body as an object of gender and sexual trouble that medical practitioners attempt to manage and contain via surgical correction. Accordingly, Chase theorizes intersex activism as a politicized struggle to achieve a broader revaluation of forms of human variation that exceed or cannot be easily categorized within the taxonomies of dimorphic sex, binary gender, and heteronormativity.

I noted above that as the millennium turned, Chase began to significantly shift her views. In her 2006 interview with Vernon Rosario that I cited at the beginning of this chapter, Chase takes care to distance her work as the founder and executive director of ISNA from feminist and queer scholarship. In contradistinction with her 1998 argument in "Hermaphrodites with Attitude," in the 2006 interview Chase claims that a radical critique of sex/gender norms and heteronormativity is neither implicit in nor particularly helpful for the intersex movement. In a remarkable departure from her earlier analysis, Chase now contends that "most intersex presentations are caused by underlying disorders."[27] This is a surprisingly different perspective on intersexuality than the feminist, queer and poststructuralist one offered in "Hermaphrodites with Attitude." Chase laments to Rosario, "I think that people in women's studies imagine that the existence of intersex people is a justification for creating a future that is radically different."[28]

Without noting that she herself had formerly staged precisely that argument, Chase goes on to express frustration with people in women's studies who use intersex as an object of evidence to support claims about the need for social transformation. The passage from the 2006 interview with Rosario in which Chase's above quotation appears reads, in its entirety, as follows.

> I think that people in women's studies imagine that the existence of intersex people is a justification for creating a future that is radically different. What I like to remind them is that intersex people have not been subjected to such an intense and harmful medicalization for very long. The ways that intersex people are treated by doctors—with shame and secrecy and unwanted genital surgeries—only became widespread in the 1960s. What that means is that there are lots of intersex people who were not treated that way. They made their way, for better or worse, in a world that was much more rigid about sex and gender

> than the one we live in today. So, radical social restructuring is not required in order for us to make the world an easier place for intersex people to live in.[29]

Chase positions "people in women's studies" as bad students of history who need to be *reminded* of the facts and rebuked for their failure to get them right in the first place. In this regard, Chase's claim that the medicalization of intersex is a relatively recent phenomenon is a curious one. Although Money's optimal gender of rearing paradigm did not become dominant in the medical establishment until the 1960s, the medicalization of hermaphroditism has a much longer genealogy in North American and European societies that dates back to the sixteenth and seventeenth centuries.[30] Moreover, while Chase refuses the reduction of intersex to an object of evidence that can be used to justify radical social restructuring, she uses the claim that the medicalization of intersex is a so-called recent occurrence as evidence for a different, more moderate claim. In so doing, Chase distorts the history of medical science, on the one hand, and repeats of the very operation she critiques, on the other. I address the former issue momentarily. Regarding the latter, Chase suggests that it is inappropriate to use intersex as evidence for arguments for radical social restructuring but appropriate to use intersex as evidence in arguments for moderate medical reform. This is a classic (neo)liberal argument applied to intersex, and it is based on a presentist political imaginary.

Yet because the meaning of intersex, like all meanings, is neither self-evident nor transparent, it remains important to ask how its meaning in any specific context is shaped by the particular claims and political desires that mobilize the category. According to Chase, despite their penchant for antiessentialism, feminist and queer theorists have been all too eager to essentialize anatomical intersexuality as a culturally subversive queer identity. Chase rejects this essentialism on (neo)liberal grounds. Subtly turning feminism's historical critique of the notion that "anatomy is destiny" back onto women's studies and queer theory, Chase remarks,

> I think that it is presumptuous to tell anyone what their identity should be. I do not think it should be part of any liberal agenda to tell anyone that their identity should be determined by their anatomy. I do not understand why that is a position that is attractive to people in queer theory and women's studies: that intersex people's identity should be determined by their anatomy.[31]

Chase suggests that "people in queer theory and women's studies" make the mistake of telling intersex people "that their identity should be determined by their anatomy" and, in so doing, betray the political philosophy of progressive liberalism, a philosophy that Chase now explicitly situates at the heart of intersex activism by invoking a "liberal agenda." Interestingly, Chase does not name which specific "people in queer theory and women's studies" she is critiquing. This generalization occludes feminist science studies critiques of intersex medicine and its complicity with a neoliberal model of subject formation.[32] Whereas Chase formerly understood intersex activism as a radical extension of queer politics and its challenges to the nationalist strictures of heteronormativity, Chase now describes intersex activism as a neoliberal movement for medical reform. In this sense, Chase's interview with Rosario reveals that her critique of the uses of intersex in women's studies and queer theory can also be read as a conservative departure from her previous argument in "Hermaphrodites with Attitude."

At the same time, as I suggested above, Chase's reconstruction of the history of the medicalization of intersex obscures the much longer genealogy of the medicalization of hermaphroditism that I glossed in my discussion of Foucault's *Abnormal* in chapter 1. Foucault suggests that debates over hermaphroditism were central to the beginnings of a clinical approach to sex in the sixteenth century and to the eventual deployment of sexuality and the emergence and expansion of biopower in the eighteenth and nineteenth centuries. Those legacies in turn shaped Money's modern biomedical approach to intersex, which pathologizes people with nonstandard anatomies so as to surgically, practically, *and* ideologically construct sexual dimorphism as a natural fact, foreclosing other classificatory schemas based on different interpretations of anatomical diversity. Chase's effort to remind people in queer theory and women's studies of the facts of history thus obfuscates the connections between the twentieth-century medicalization of intersex and the longer history of the medical regulation of hermaphroditism, sex, and sexuality more generally.

In other words, the claim that medicalization is a recent occurrence, like the activist focus on surgeries alone, produces an ahistorical view of medicalization. Chase's argument is significant not only as an example of single-issue neoliberal politics, but also because it denies a genealogical view of surgery (a la Foucault) as a key component of the modern technologies of sexual subjectivation that produce the idea of subjects with "insides" to be seen, subjects that hide the secrets of sex within.[33] Surgeries aren't just one technology among many—they are central to sexual subjectivation in modernity and therefore inextricably connected to sex/gender and sexual relations.[34]

Social Construction Theory, Standpoint Epistemology, and the Politics of Intersex Experience

In the interview with Rosario, Chase stresses that it is not gender, but rather medicalization, defined as "unwanted genital surgeries," that is at the root of the problems intersex people face. Alice D. Dreger and April M. Herndon provide some useful context for Chase's argument.[35] Dreger and Herndon argue that Money's optimal gender of rearing (OGR) model advocated early surgical intervention on the basis of a social constructionist hypothesis: that the infant's gender is more or less plastic until roughly eighteen months after birth. As I discussed in chapter 1, Money thought that a combination of surgery and hormonal treatments could normalize the infant's sex, giving the infant the basis on which to form a "normal" gender identity. In turn, some feminists (such as Ann Oakley, whose work I also examined in chapter 1) appropriated the concept of gender from Money and refashioned it as a critical tool for challenging biological determinist arguments about the nature of sexual inequality. In a particularly striking passage, Dreger and Herndon refer this history back to Chase:

> Chase has argued that it is the very obsession with "the gender question" that has led to so much harm for people with intersex. According to Chase, while some people (like Money and some feminists) have used intersex to sit around debating nature versus nurture, real people with intersex have been hurt by these theories and their manifestations. Chase has therefore argued that "intersex [has been] primarily a problem of stigma and trauma, not gender."[36]

Chase's claim that intersex is not ultimately about gender disclaims and contradicts her earlier analysis of intersexuality in "Hermaphrodites with Attitude." It also denies the history of intersex as the concept that gives us gender, as my genealogy in chapter 1 proposes. In my view, it is unfair to characterize feminist analyses of intersex as reducible to armchair theorizing about nature versus nurture, to suggest that feminist scholarship is inattentive to the material effects of knowledge production, and to haphazardly group "Money and some feminists" together, as if there are no qualitative differences between biomedicine and feminism. Nonetheless, I agree with Dreger and Herndon's reading of Chase in at least one respect, that it is the obsession with gender and sex *normativity* that transforms intersex into a medical problem.

In 2006, when she was a member of ISNA, April Herndon posted an essay on the ISNA website titled "Why Doesn't ISNA Want to Eradicate Gender?" There, Herndon explained ISNA's then-current position on gender in further detail. She is worth quoting at length:

> Intersex people don't tell us that the very concept of gender is oppressive to them. Instead, it's the childhood surgeries performed on them and the accompanying lies and shame that are problematic.
>
> Again, many of these surgeries are performed with the belief that these procedures will help a child settle into a gendered world, but that doesn't mean the whole system of gender must fall in order for people with intersex conditions to live happy, fulfilling lives. It simply means that these surgeries and the shame that surrounds them are an unfortunate instantiation of problematic gender norms and we should work on ending unwanted surgeries and stigma.
>
> There are, of course, some people with intersex conditions who identify as a third gender or gender queer—just as there are some people with completely typical sex anatomies who don't identify as strictly male or female. Our aim at ISNA isn't to undermine these people's goals, or to suggest that people who identify as a third gender don't exist or don't matter, or to suggest that everyone must adopt a gender. Rather, we hope to end painful and unnecessary childhood surgeries that rob people of corporeal autonomy and sexual function because everyone—regardless of gender identity—deserves that. And we hope to end the shame and secrecy that cause so much pain for so many people with intersex conditions.
>
> We hope that scholars, particularly those invested in helping members of marginalized groups gain a voice in conversations about themselves, will take seriously the concerns about surgery, secrecy, and shame raised by intersex people and understand that ISNA and the majority of its constituency don't necessarily share the goal of eradicating the very notion of gender.[37]

Herndon's comments take as their starting point the experiences and voices of intersex people. Validating these marginalized viewpoints, Herndon, like Chase, implicitly privileges liberalism—and its emphasis on individual autonomy and the right to define oneself free from the interference of others—as the defining political philosophy of ISNA's activism. Moreover, Herndon repeatedly stresses the motif that it is "childhood surgeries . . . and the accompanying lies and shame" that cause "so much pain for so many people with intersex conditions." In this regard, she almost seems to define intersex activism in terms of a single-issue agenda, that of ending infant genital surgeries.[38] Explaining INSA's specific goals, Herndon underscores what she sees as the fundamental differences between ISNA's concerns with childhood surgeries, lies, and shame versus the concerns of "those who identify as a third gender or gender queer" or "share the goal of eradicating the very notion of gender." By juxtaposing ISNA's project with genderqueer activism, Herndon makes it appear as though questioning the relation between the medicalization of intersex and processes of gender regulation is ethically inappropriate, a betrayal of what the intersex people ISNA represents feel and experience. Thus, Herndon's liberalism combines with an appeal to experience as the ground of truth about intersex life.

In centering the voices and experiences of intersex people, Herndon's account converges with feminist standpoint epistemology, which posits subordinated groups' experiences as a basis of legitimate knowledge about systems of oppression.[39] However, feminist standpoint theories also encourage us to analyze minoritized subjects' experiences and narratives critically, rather than simply taking them at face value, so as to avoid the reification of dominant ideological systems, which may sometimes be internalized by oppressed peoples. As Patricia Hill Collins puts it, standpoint epistemology emerges out of conscious struggles "to develop new interpretations of familiar realities."[40]

Challenging the stability and coherence of experience as a ground of knowing, Joan W. Scott has argued that to construe experience as a transparent, unmediated source of truth about the world is to risk reifying, rather than disarming and displacing, entrenched ideological systems.[41] This does not mean that experience does not have an important role to play in the production of situated knowledges and strategies for resisting and transforming oppressive structures. But it does mean that the *construction* of experience requires critical analysis.

> When experience is taken as the origin of knowledge, the vision of the individual subject (the person who had the experience or the historian who recounts it) becomes the bedrock of evidence on which explanation is built. Questions about the constructed nature of experience, about how subjects are constituted as different in the first place, about how one's vision is structured—about language (or discourse) and history—are

> left aside. The evidence of experience becomes evidence for the fact of difference, rather than a way of exploring how difference is established, how it operates, how and in what ways it constitutes subjects who see and act in the world.
>
> To put it another way, the evidence of experience, whether conceived through a metaphor of visibility or in any other way that takes meaning as transparent, reproduces rather than contests given ideological systems.[42]

The problem with the uncritical acceptance of the self-evidence of experience becomes evident in the way that Herndon moves from the claim that "intersex people don't tell us that the very concept of gender is oppressive to them" to the conclusion that gender is irrelevant to intersex experience and political activism. Thus, even as she makes the important point that scholars interested in aiding marginalized communities need to listen carefully to those communities' concerns, Herndon also seems to assume that listening carefully means not hearing gender at all—a proposition that is called into question by Herndon's own recognition that "these surgeries and the shame that surrounds them are an . . . instantiation of problematic gender norms." Ending unnecessary surgeries without disarming and displacing those problematic gender norms may therefore be a problematic strategy in light of the genealogy I traced in chapter 1.

Herndon's appeal to experience is a common move. It provides a way for minoritized subjects to contest accounts of the world that exclude or marginalize them. But, as Scott argues, the appeal to experience often relies on the very epistemological presumptions that naturalize reigning ideological systems. This does not mean experience is unimportant, but it does mean that its status as evidence needs to be interrogated and historicized, both by marginalized subjects and by scholars. This point becomes clear in the 2002 essay by Emi Koyama and Lisa Weasel titled "From Social Construction to Social Justice: Transforming How We Teach about Intersexuality," mentioned at the outset of this chapter. Koyama and Weasel's essay not only provides further background to Chase's, Dreger and Herndon's, and Herndon's arguments, but also highlights the dangers of dualistic modes of thought that privilege the truth of experience over and against theory and critical analysis.

Koyama and Weasel suggest that, by treating intersex as evidence of the social construction of sex/gender, women's studies scholars obscure the real-world concerns and experiences of intersex persons. They argue that women's studies scholars need to "recognize that it is not the responsibility of intersex people to deconstruct binary gender-sex or to be used as guinea pigs to test out the latest theories about gender." "Do not be disappointed," they continue, echoing the claims of Chase and Herndon, "that many intersex people are not interested in becoming members of the third gender or overthrowing sex categories altogether."[43] The problem with the way intersex is taught in women's studies, Koyama and Weasel conclude, is that "intersex existence is understood and presented largely as a scholarly object to be studied in order to deconstruct the notion of binary sexes (and thus sexism and homophobia) rather than as a subject that has real-world implications for real people."[44]

Koyama and Weasel's language reveals the centrality of the theory/real world binary to this and other experience-based critiques of academic feminism and women's studies. They insist repeatedly that intellectual abstraction in general and social construction theory in particular are a priori incapable of speaking to or shedding light on issues that have "real-world implications for real people." This premise reproduces a reductive opposition between the "socially constructed" and the "real," wherein the former somehow exists independently of, and has no relation to, the latter. This binary in fact undercuts Koyama and Weasel's own argument, which ultimately seeks to hold women's studies scholarship and teaching accountable for the objectification of intersex individuals. Indeed, they blame women's studies for using intersex people as "guinea pigs," thus drawing on a metaphor from the world of science and biomedicine in order to subtly shift and displace the blame from science to women's studies.

In addition, it is the theory/real world binary that allows Koyama and Weasel to suggest that accounts of intersexuality that attend to the workings of "binary sexes (and thus sexism and homophobia)" lack real-world relevance. This claim presupposes a contradictory and ultimately untenable understanding of theory's relation to sociopolitical life. Koyama and Weasel seem to assume that theory exists independently of the social and political conditions in which it is produced even as they argue that theory can also mystify those conditions, and they also assume that binary sex, sexism, and homophobia are purely theoretical constructs. For reasons they never openly state, Koyama and Weasel hold that these terms are incapable of referring to real, material forces and forms of power that shape people's lives, especially the lives of intersex people. Moreover, Koyama and Weasel never address the problem of how to get at experience. They treat experience as an unmediated category, as if people's claims give transparent access to the real.

By contrast, feminist thinkers like Butler, Haraway, Collins, Spivak, and Scott argue that experience does not exist independently of the social, political, and economic conditions in which it is produced. Those material conditions are themselves constituted, in part, through implicit and explicit theoretical and ideological practices and commitments that are never transparent, which is why they require critical interrogation. Feminists have long argued for the need to name and analyze the ways in which power relations privilege some groups at the expense of others.[45] In this understanding, theory is a vehicle for naming and analyzing how material-semiotic forces such as sexism, racism, and homophobia operate in the world.[46] From this perspective, Koyama and Weasel's argument for replacing the emphasis on "deconstruction" and "social construction" in women's studies with an emphasis on social justice flattens out processes of social formation and obscures the importance of attending to contestations over how to define and implement social justice. We have to reckon with the possibility that different subjects are produced by different histories and that they may have different visions of justice.[47] To oppose "theory" to the "real world" or "practice" is to reify an illusory and anti-intellectual distinction between realms of praxis (broadly speaking, thought and action) that are in fact inextricably connected. The point is that what gets to count as *theory* and as *practice* is overdetermined by social, cultural, and political economic relations. As I suggested in the previous chapter, feminist theories of the gendered regulation of intersex bodies, such as those of Kessler, Fausto-Sterling, and Butler, are valuable precisely because they speak to the relays between theory and practice that inform the biopolitics of embodiment. By critically attending to such relays, it becomes possible to question the naturalization of the authority of institutions such as science and biomedicine, precisely those institutions that, unlike women's studies, literally use and experiment on guinea pigs—not only to produce knowledge but also to reproduce hierarchies that determine which bodies and life forms get to count as normative, legible, healthy, and valuable, and which do not.

Intersex and Gender

I now want to turn to the claim, made above in different forms by Chase, Dreger and Herndon, Herndon, and Koyama and Weasel, that gender is, in the final instance, peripheral to intersex issues. This claim also appears in Vernon A. Rosario's 2007 essay "The History of Aphallia and the Intersexual Challenge to Sex/Gender," which explores the ways in which the history of the medical treatment of hermaphroditism in previous centuries informs current treatment standards for intersex and extends Koyama and Weasel's argument that gender theory obfuscates the material realities and concerns of individuals with intersex conditions.[48] Rosario notes that "the mainstream intersex support groups centered around particular diagnoses (such as androgen insensitivity syndrome, hypospadias, or congenital adrenal hyperplasia) have intensely debated if not completely rejected the intersex label because the affected individuals feel their gender identity is either male or female and they do not want to be perceived as gender intermediates."[49] Rosario makes a similar point in his 2004 essay "The Biology of Gender and the Construction of Sex?"[50] There, Rosario defines intersexuality as an "umbrella medical term"[51] for a variety of "objective, material conditions, not indications of an elective gender identity."[52] "Despite discordant sex chromosomes, genitals, and/or gonads," Rosario writes, "the vast majority of intersexed people have a definite gender identity as male or female; they are not inter-gendered."[53] The certainty with which Rosario distinguishes the "objective" biological aspects of intersexuality from the "merely cultural" or social aspects of male and female gender identities is striking. Is a larger-than-standard clitoris an objectively measurable feature of certain bodies or is it, rather, a cultural interpretation that not only presumes but also institutes a corporeal norm? Indeed, I would suggest that the difference between biological intersex conditions and social gender identity can only seem self-evident when we presume that the biological dimensions of life are not imbued with symbolic value. However, as Morgan Holmes argues, "the biological sciences and biomedicine do not produce views of organic function or metaphors for embodiment apart from a cultural setting that provides the very possibility for their meaning."[54] Rosario's argument suggests that bodies do not have a history when in fact, as I proposed in the previous chapters of this book, the history of the medicalization of intersex can be understood as a history of bodily regulation through the terms of gender and other intersectional categories of difference including race, class, and nation (a claim on which I further expand in the next two chapters).

As Holmes suggests, it is important to think carefully about how cultural settings shape the meanings attributed to intersex bodies, because those meanings have material consequences. While Rosario is indeed correct that many affected parties and groups have rejected the intersex label and that many people with intersex conditions identify as male or female, his analysis overlooks the ways in which gender informs the medicalization and codification of intersex as a set of "particular diagnoses." This is not to say that these diagnoses are unreal or immaterial, far from it. But it is to suggest that it is important to ask how the materiality of intersex conditions is shaped, in part, by gendered logics.

When physicians diagnose an infant with an intersex condition, for instance, when they consider an infant's genitalia to be atypical enough to require surgical intervention, do not normative gender assumptions about proper genital shape, size, and function inform such diagnoses and treatments? Furthermore, isn't it the case that many physicians and parents presume that "normal"-looking genitals, whether inborn or surgically produced, are a precondition for normative and legible forms of gendered and sexual personhood? And what are we to make of the fact that intersex gave birth to gender not merely conceptually but in the establishment of a history of treatment?

In "The History of Aphallia and the Intersexual Challenge to Sex/Gender," Rosario situates his argument as an endorsement of ISNA's agenda to end unnecessary infant genital surgeries and to make complete medical disclosure and patient and family counseling services central to future strategies for the medical management of intersexuality. But "ISNA," Rosario writes, "has struggled to shift the focus of intersex politics from sex/gender theory battles to practical clinical concerns."[55] According to Rosario, this is not only because medical and media attention has tended to sensationalize intersexuality as a "gender crisis," but also because feminist and queer theorists have used intersex predominantly "as an opportunity for destabilizing biological notions of sex and gender."[56] Rosario thereby suggests that "sex/gender theory battles" have gotten in the way of, or deflected much-needed attention from, the project of reforming the medical standards of intersex management.

For this reason, Rosario concludes "The History of Aphallia and the Intersexual Challenge to Sex/Gender" by singling out feminist and queer theorists for critique, stating bluntly that,

> while the historical construction of sex, gender, and hermaphroditism certainly inform the current "optimal gender" paradigm of treatment, deconstructing these will not make intersexuality disappear any more than it will erase the categories of sex and gender. No amount of theorizing about intersex or its cultural impact on gender theory will eliminate the physical pain, infertility, endocrinological disorders, and emotional stress that burden many people with intersex conditions.[57]

In this passage, Rosario seeks to drive home the point that intersex refers to real, material, physiological phenomena, not a discursive fiction that can be rendered immaterial by the theoretical magic of "deconstruction," a term that Rosario, like Koyama and Weasel, uses as if it were almost a slur.[58] Contra its precise, technical meaning, for Rosario to "deconstruct" means to make "disappear" or to "erase." Rosario paints deconstruction as a purely negative activity, when in fact a more nuanced understanding of deconstruction grasps its affirmative as well as critical aspects.[59] Conflating "deconstruction" with the project of gender theory in general, Rosario elaborates a stark division between "theorizing about intersex" and the materiality of intersex bodies and experiences. He invokes the "physical pain, infertility, endocrinological disorders, and emotional stress that burden many people with intersex conditions" in a manner that suggests that these sensory and bodily experiences are at once prior to and fundamentally separate from gender as it is lived, theorized, and regulated. While Rosario is surely right that theorizing about intersexuality or its cultural impact on gender theory cannot in any immediate way alleviate the physical pain, infertility, endocrinological disorders, and emotional stress that many intersex persons do indeed face, it is unclear why he needs to cast gender theory as purely discursive and out of touch with reality to buttress this claim. For Rosario, it would seem that much gender theory is premised on a delusion: that "deconstructing" categories will lead to their disappearance and erasure. However, it is important to ask whether this is an accurate description of the project of gender theory, a point I follow up on below. It is also important to ask whom or what Rosario and Koyama and Weasel are implicitly targeting in their critiques of "deconstruction."[60]

Another influential scholar who has addressed the tension between intersex activism and feminist scholarship is Katrina Karkazis. Her 2008 monograph *Fixing Sex: Intersex, Medical Authority, and Lived Experience* provides one of the most comprehensive evaluations to date of controversies surrounding the medical management of intersexuality in the United States. Karkazis casts the tension in a slightly different light, in terms of a fundamental incongruence:

> Although in many respects activists' analyses of treatment for intersexuality draw on theories about the social construction of gender—and intersexuality has been used as a prime example of this phenomenon in women's studies and queer studies classes—the two fields' interests are not congruent. Most feminist interest in intersexuality stems from its value as a heuristic device, but the goal of many people with intersex conditions and activists is not to deconstruct or eliminate gender, or to advocate for a third sex or no sex, but rather to change treatment practices and improve the well being of others with these conditions. While this goal necessarily involves a broadened understanding of what it means to be male or female, the cornerstone of this argument does not center on gender: intersexuality, these

activists argue, is primarily a problem of stigma and trauma, not of gender.[61]

Though Karkazis argues throughout *Fixing Sex* that "the lens of gender literally shapes the body" and that this shaping is particularly evident in the case of "individuals who undergo treatment procedures and interventions for intersexuality,"[62] in the above passage, which appears in the conclusion to her book, Karkazis affirms the activist view that intersex is not primarily a problem of gender. In fact, the last sentence of the quotation is taken verbatim from ISNA's mission statement: "Intersexuality is primarily a problem of stigma and trauma, not gender." ISNA began to assert this claim on its website in the mid-1990s, adding, among other points, that "parents' distress must not be treated by surgery on the child" and that "all children should be assigned as boy or girl, without early surgery."[63] In paraphrasing ISNA's argument, Karkazis fulfills one of the main goals of her project, which is to use standpoint theory to validate the lived experiences of intersex persons *against* the dominant medical narrative of intersexuality. Yet by corroborating ISNA's position, Karkazis misses out on an important opportunity to critically engage the assumptions underpinning both ISNA's analysis and ISNA's use of lived experience in that analysis.[64]

I now want to turn directly to ISNA's claim that "intersexuality is primarily a problem of stigma and trauma, not gender" to examine the claim's underlying assumptions. If "intersexuality is primarily a problem of stigma and trauma, not gender," then why does ISNA argue for the need to assign "all children as boy or girl," albeit with the important qualification, "without early surgery?" In short, if the problem of intersexuality is not gender but surgical sex assignment, then why must the answer to intersexuality still involve gender assignment?

As Noah Ben-Asher astutely observes, Butler has recently implied but ultimately chosen to dismiss such a critique of intersex politics. Querying the relation between queer theory and the intersex movement, Butler writes,

> It does not follow, therefore, that queer theory would oppose all gender assignment or cast doubt on the desires of those who wish to secure such assignments for intersex children . . . [T]he perfectly reasonable assumption here is that children do not need to take on the burden of being heroes for a movement without first assenting to such a role. In this sense, categorization has its place and cannot be reduced to forms of anatomical essentialism.[65]

According to Butler, because nearly all children are raised as one of two normative genders, it would be unethical to force some children to be "heroes for a movement" by raising them without a gender or as a third gender. "Although Butler's position makes intuitive sense," Ben-Asher writes, "it relies on difficult reasoning."[66] "It seems," Ben-Asher continues,

> that Butler views the possibility of not assigning a gender to a child (or assigning a third or an intersex gender) as a possible source of social ostracism. But this same reasoning is used by John Money and others to justify intersex surgery: the child will adjust better to the environment with "normal" looking genitals than with genitals that are unintelligible. Therefore, challenging sex assignment while using the same logic to justify gender assignment deserves rethinking.[67]

By pointing out that the activist argument for the necessity of normative gender assignment relies on the "same logic" both Butler and ISNA find problematic in surgical sex assignment, Ben-Asher reveals that gender assignment, like normalizing genital surgery, may well also violate the liberal autonomy that Chase and Herndon invoke as the justification for their arguments against surgical intervention. In short, gender assignment too engenders bioethical and biopolitical concerns about the norms and regulations imposed on children's bodily integrity. The point is not that the intersex critique of sex assignment is wrong, but rather that it perhaps does not go far enough. What forms of power underlie the pervasive insistence on the imperative to gender subjects? What has gender become such that it repeatedly gets posited as a precondition without which human life as such is said to be unthinkable and unlivable? Building on Ben-Asher, it is my contention that liberal humanism need not be the only resource we turn to as we contemplate these questions. For instance, Foucault provides a powerful way of thinking about technologies of the self and the erotic as practices of freedom.[68] Using those Foucauldian resources, future research might, for instance, engage intersex embodiment in terms of how the erotic sensorium is regulated, sometimes drastically compromised and enervated, but also sometimes capacitated and invigorated by processes of biomedical normalization.

In addition, the claim that "intersexuality is primarily a problem of stigma and trauma, not gender" presumes that stigma and trauma are clearly separable from, and have no causal or correlative relation to, gender as a multidimensional structure of power. Considered generally, for example, in terms of contemporary social relations between men and women, this presumption would appear difficult to sustain. As diverse thinkers including Butler, Stone, Wolf, Collins, Kimmel, Halberstam, Ferguson, and many others have argued, normative gendered bodies, behaviors, and identities are constructed in and through the policing and regulation of what are perceived to be gender nonconforming bodies, behaviors, and populations.[69] Effeminate boys, tomboys, queer men, butch lesbians, masculine women, cross-dressers, trans people, genderqueer folk, people of color, and especially trans women of color are often regarded as "gender outlaws," to use Kate Bornstein's term.[70] These folks are recurrently ostracized, stigmatized, and subjected to violence or the threat of violence for breaking with hegemonic regimes of gender normativity as they are shaped by whiteness, nationalism, and xenophobia.

These gender regimes are also frequently heteronormative. As Dreger has shown in her history of the medicalization of hermaphroditism during the nineteenth and twentieth centuries, "a significant motivation for the biomedical treatment of hermaphrodites is the desire to keep people straight."[71] Dreger means that phrase—"keeping people straight"—both literally and figuratively. As she puts it, "Many assume that if we don't keep males and females sorted, social institutions that we hold dear—including divisions into heterosexuality and homosexuality, into mothers and fathers, into women athletes and men athletes—will no longer be viable."[72] Hermaphroditism was viewed by British and French medical specialists starting in the late 1800s as a threat to the social order, in particular, as a threat to naturalized sexual divisions and hierarchies of labor, identity, and sociality. To keep that threat at bay, twentieth-century western biomedical experts developed a range of surgical and hormonal techniques to "normalize" infants born with intersex conditions. In this sense, the stigma and trauma persons with intersex embodiments face, including unwanted genital surgeries, hormone treatments, repeated medical inspections, and family shame and secrecy, are fundamentally related to gender and sexuality as regulatory structures.

In other words, ISNA's argument that "intersexuality is primarily a problem of stigma and trauma, not gender" performs what Gayatri Spivak calls, in literary critical terms, a metalepsis: the "substitution of one figure for another," in this case, the substitution of "effect for cause."[73] ISNA's metalepsis may be a strategic political move (a strategy designed to deflect attention from broader considerations of gender and to focus squarely on the problem of normalizing infant genital surgery). However, the strategic function of the metalepsis doesn't contravene the fact that individuals with intersex features are stigmatized and traumatized neither by accident nor randomly, but rather for a very particular set of historical reasons. They are stigmatized and traumatized because western biomedicine and hegemonic legal and social institutions are deeply invested in maintaining the ideologies of sexual dimorphism and binary gender.

To understand that intersex is not only a problem of stigma, trauma, and unwanted genital surgeries but *also* and *simultaneously* of gender is therefore to begin to grasp the value of what Salamon, cited at the beginning of this chapter, calls "the systematic understanding that women's studies provides of the structures of gender—and the relations of power that underlie those structures." From this perspective, it is significant that Chase, Dreger and Herndon, Herndon, Koyama and Weasel, Rosario, and Karkazis all elaborate the meaning, parameters, and goals of intersex activism not only against the dominant medical view of intersexuality, but also, and crucially, *against* what they variously call women's studies, queer theory, feminist scholarship, and gender theory. Implicit in their critiques is the presumption that these field formations are dominated *en total* by what their authors variously label the desire: for "radical social restructuring" (Chase), to "eradicate the very notion of gender" (Herndon), to "overthrow sex categories altogether" (Koyama and Weasel), to "make intersexuality disappear . . . [and] erase the categories of sex and gender" (Rosario), and to "deconstruct or eliminate gender, or to advocate for a third sex or no sex" (Karkazis). Taken together, these characterizations paint women's studies, queer theory, feminist scholarship, and gender theory as sites of profound negativity and aggression, and, worse still, as academic fringes united around some ridiculous and senseless hostility toward the very ideas of gender and of sex.

This portrayal is unfairly reductive and homogenizing, as a great deal of research in feminist and queer studies differs from and challenges these authors' characterizations in important ways. I would even venture to say that most feminist and queer scholars are not expressly interested in eradicating or eliminating gender altogether, but rather are concerned with identifying, analyzing, and challenging gendered and related forms of *inequality* and *oppression*. Much feminist and queer scholarship produced during the last three decades focuses on the interdependent and co-constitutive dynamics of structures of subjugation such that sexism, racism, classism, ableism, transphobia, and xenophobia are theorized as fundamentally intersectional. Analyzing these intersections creates opportunities to forge coalitional strategies of resistance and social transformation. Other important strains of feminist and queer scholarship engage postcolonial and indigenous studies, transnational perspectives, Marxism, and new materialism to rethink the intersections of gender oppression and heteronormativity with settler colonialism, state violence, white supremacy, westocentrism, and speciesism. In a related but distinct vein, one especially relevant to this book's argument, queer feminist science studies offers interdisciplinary accounts of how science, medicine, and technology shape and are shaped by ideologies of sex, gender, race, ability, and sexuality.[71] As this brief discussion suggests, feminist and queer scholarship is heterogeneous. It encompasses a truly rich and diverse array of analytic and interpretive approaches, epistemologies, and political desires. Yet few feminist and queer scholars, so far as I am aware, propose that the answer to the problems they address is simply to get rid of gender and/or sex altogether, as if that were even possible in the current conjuncture.

Their reductionism notwithstanding, and on the other hand, what remains revealing about Chase's, Dreger and Herndon's, Koyama and Weasel's, Rosario's, and Karkazis's analyses, is the negativity and aggression they locate in feminism and queer theory; and, more importantly, the negativity and aggression they expose as central to intersex politics, which they in turn enact by pitting intersex politics against feminism and queer theory to legitimize the former at the expense of the latter. The place of negativity in politics has been the subject of much recent critical debate.[75] According to Elizabeth Wilson's account, negativity is "intrinsic (rather than antagonistic) to sociality and subjectivity," which means that, "in important, unavoidable ways, feminist politics attack and damage the things they love."[76] Wilson's insight is not exclusive or unique to feminist politics. All forms of politics enact various explicit and implicit aggressions toward the things they love as well as toward those they mark as other. Politics materializes, after all, in and through the establishment of the distinction between friend and enemy.[77] If Wilson is right, then future research might more thoroughly investigate the place of negativity in intersex politics. What if intersex politics is not only reparative and beneficent but also produces harms of its own that are hard to bear?

As Ben-Asher points out, "there is an ethical concern when group projects of de-subjugation undercut each other."[78] Thus, even as these authors highlight the importance of thinking through the ethics and politics of the uses of intersex in feminist and queer work, so too do their writings inadvertently draw attention to the ethics and politics of the uses of feminism and queer theory in intersex politics. What kinds of political work get done, what kinds of alliances are enabled and foreclosed, when the meaning of intersex activism is defined by delegitimizing women's studies and queer theory? What if, rather than understanding intersex activism and feminist and queer scholarship in oppositional and mutually exclusive terms, we rethought them as critical interlocutors with nonidentical interests—as was once the case when Chase began dialoguing with Kessler and Fausto-Sterling and founded ISNA?

Conclusion

In this chapter I have tried to underscore the ways in which intersex critiques of women's studies and queer theory end up confirming the intellectual relevance of the very objects they renounce. Nowhere is this more evident than in ISNA's central claim: that "intersexuality is primarily a problem of stigma and trauma, not gender," which, as I suggested above, disavows the very structure (heteronormative gender) that sustains the cultural norms that hold that intersex bodies are shameful and sources of stigma and trauma and therefore require nonconsensual surgeries in the first place. Indeed, ISNA's claim that "intersexuality is primarily a problem of stigma and trauma, not gender" is only possible in the context of a field—women's studies—that helped put gender on the map in ways that exceed and trouble the concept's originary codification as a technology of biomedical regulation in Money's OGR paradigm. While feminist and queer critical engagements with intersex cannot alone produce biomedical reforms or institutionalize legal protections for intersex people (only large-scale democratic agitation can achieve those ends), these bodies of thought do offer a set of critical perspectives that enable us to rethink the politics and ethics of sex and gender regulation in new ways. As Chase argued in "Hermaphrodites with Attitude," such critical perspectives can enliven, enrich, and transform the scope of intersex politics.

Despite the concerns I have raised about the routinization of intersex critiques of feminist and queer scholarship, I ultimately agree with Chase, Dreger and Herndon, Koyama and Weasel, Karkazis, and Rosario that debates over topics like nature and nurture have real-world consequences. These consequences differ markedly depending on whether or not one is medically diagnosed and treated as intersex, trans, or gender nonconforming. But rather than seeing these consequences as cause for shutting down debate and writing off the intellectual value of feminist and queer thought, I contend that we need to do precisely the opposite. We need to encourage debate, critically contest received ideas, and affirm and proliferate the uncertainties that we find most difficult to acknowledge.

4

Provincializing Intersex

Transnational Intersex Activism, Human Rights, and Body Politics

> Do we really need to change some children to make them human enough to get human rights?
>
> —Alice Dreger[1]

How are debates about intersex shaped by the politics of difference and struggles for sexual and gender justice in a multicultural, transnational world? How do activist and academic critiques of the medicalization of "bodies in doubt" rearticulate the meaning and materiality of human rights in a neoliberal landscape?[2] How do geopolitics, colonial legacies, consumer citizenship, and biopower inform and mitigate contemporary intersex politics? And how might transnational feminist perspectives contribute to a critical rethinking of the local and global travels and trajectories of the intersex movement?

In the space opened up by these questions, this chapter explores how US debates about intersex are shaped, challenged, and interrupted by global activism and transnational feminist perspectives.[3] Acknowledging that the term "transnational feminism" is contested,[4] I employ it here as "an intersectional set of understandings, tools, and practices" that can attend "to racialized, classed, masculinized, and heteronormative logics and practices of globalization and capitalist patriarchies, and the multiple ways in which they (re)structure colonial and neocolonial relations of domination and subordination."[5] Transnational feminisms both draw from and are aligned with other traditions of feminist praxis (including postcolonial, women of color, indigenous, materialist, queer, and poststructuralist feminisms). Challenging "monological and monocausal approaches to subjectivity and power"[6] and refusing "colonial logics of similitude,"[7] they do not presume that all sexed/gendered subjects around the globe are essentially the same, or that sex/gender is separable from other axes of difference.[8] Transnational feminist perspectives are therefore distinct from, and adopt a critical stance toward, cosmopolitan and internationalist celebrations of "global sisterhood." Furthermore, they do not view capitalist globalization's cross-border flows (of people, goods, capital, and information) as either inherently liberatory or exclusively exploitative. Instead, transnational feminist analytics focalize the shifting, unstable, but vital interdependencies between nation-states, political economies, social formations, and subjects—revealing these entities to be fundamentally contested, non-natural, nonidentical across space and time, and laced with contradictions.[9] Recognizing that "there is no such thing as a feminism free of asymmetrical power relations," transnational feminisms "involve forms of alliance, subversion and complicity within which asymmetries and inequalities can be critiqued."[10]

I use transnational feminist perspectives to argue that US-based intersex advocacy risks reiterating structures of US and global northern dominance when it does not self-reflexively interrogate its own politics of location in relation to transnational histories of imperialism, neoliberalism, and biopower. In her reading of Adrienne Rich's influential concept of the "politics of location," Caren Kaplan argues that practicing ethical accountability in transnational contexts requires acknowledging "the historical roles of mediation, betrayal, and alliance in the relationships between" various subjects in diverse locations.[11] Drawing on Kaplan's intervention, I investigate how human rights discourse, colonial legacies, biopolitics, and neoliberal ideologies contour the locational politics of US intersex activism. I do so by examining two crucial events in the history of the Intersex Society of North America (ISNA), an advocacy group that became, during the tenure of its existence (1993–2008), the most highly visible and influential intersex activist organization in the world: 1) ISNA's failed attempt to lobby for the inclusion of intersex surgery in the US Congress's 1997 federal ban on "female genital mutilation" (FGM); and 2) ISNA's influence on two 1999 decisions by the Constitutional Court of Colombia to rework the definition of informed consent and to limit doctors' capacity to perform normalizing genital surgery.

I contend that transnational feminist perspectives provide indispensable tools for analyzing the politics of location of US intersex activism and, ultimately, for provincializing and decolonizing the intersex imaginary. I use the term "intersex imaginary" to refer to shared yet situated ways of imagining intersex bodies.[12] The intersex imaginary is a site of political struggle and contestation. US and western understandings of intersex are historically and geopolitically particular, not universal. In the English language, for instance, the term *intersex* assumes an analytic separation between sex, gender, and sexual orientation that is not indigenous to many cultures around the globe.[13] For this reason, articulations of the intersex imaginary can falsely universalize western narratives of intersex consciousness, personhood, and political organizing as paradigmatic, misrecognizing different configurations of embodiment and subjectivity as nascent reflections of a universal intersex identity. The potential imperialism of constructing the intersex imaginary in US/Eurocentric ways can be countered by provincializing that imaginary, a project with decolonial motivations.[14] This project begins with the recognition that colonialism lives on in the history of the present in myriad ways: in the processes of white settler colonialism that established the US government's territorial control of native lands and resources; in the uneven and unequal distribution of life chances under globalization; and in ongoing practices of US and western political and economic intervention in international affairs. Provincializing the intersex imaginary involves connecting ways of thinking about intersex bodies to these histories—in short, to specific times and places. It entails critiquing the geopolitical overdetermination of US and western understandings of intersex as they circulate globally and are transformed and sometimes called into question in the process. The project of provincializing intersex reveals that the imprinting of place on conceptions of intersex embodiment has crucial but underinterrogated implications for transnational struggles for corporeal freedom.

The first section of this chapter tracks the global rise of the concept of intersex human rights. The second section synthesizes transnational feminist perspectives on human rights. The third section analyzes ISNA's lobbying efforts around intersex surgery and FGM. And the fourth section closely reads ISNA's amicus brief for the Colombian Constitutional Court, the court's decisions themselves, and critiques of those decisions by scholars and activists located in Colombia and abroad. Through these analyses, I show that transnational feminist perspectives offer new ways of understanding the local and global effects and implications of US intersex activism. Concurrently, I argue that the transnational regulation of sexed bodies occurs not only through the globalization of western biomedical conceptions of sex/gender normativity, but also through global circulations of human rights discourse and impositions of US neoliberal democratic frames of subjectivity. My reading of the Colombian case focuses on how these frames travel across national borders and solidify, even as they are challenged and interrupted by local actors. Both ISNA's FGM analogy and the organization's influence on the Colombian court's decisions suggest that provincializing intersex requires rethinking the imbrication of US intersex activism in western imperial formations and neoliberal logics.

Intersex Human Rights

Over the past two decades, intersex became a "human rights issue," as one ISNA press release put it in 2005.[15] The press release asserts "that the standard medical approach to intersex conditions leads pediatric specialists to violate their patients' human rights."[16] Cheryl Chase, the longtime executive director of ISNA, is quoted in the press release as saying that intersex people

> deserve the same basic human rights as others . . . No longer should we be lied to, displayed, be injected with hormones for questionable purposes, and have our genitals cut to alleviate the anxieties of parents and doctors. Doctors' good intentions are not enough. Practices must now change.[17]

Chase's impassioned remarks foreground the high stakes of contestations over the medical treatment of intersex people. They also prefigure a handful of contemporary events from around the globe that highlight the growing transnational circulation of the notion of intersex human rights.

According to Julie A. Greenberg, legal institutions have the potential to play a significant role in protecting the rights of intersex people.[18] As I discuss below in greater detail, in 1999, Colombia became the first country in the world to use human rights as a ground for imposing legal regulations on intersex genital surgery. In 2000, in recognition of ISNA's influence (on the Constitutional Court of Colombia's decisions and in other arenas), the International Lesbian and Gay Human Rights Commission awarded ISNA the Felipa de Souza Award for making "significant contributions toward securing the human rights and freedoms of sexual minorities anywhere in the world."[19] In the years following, international awareness about and media interest in intersex has continued to increase, thanks in large part to the work of ISNA and other intersex alliances.

In 2005 and 2011, respectively, the San Francisco Human Rights Commission and the European Network of Legal Experts published critical reports on the medical normalization and social stigmatization of intersex people.[20] In 2013, highly visible international figures and associations, including the UN Special Rapporteur on Torture, the Inter-American Commission on Human Rights, the Australian Senate, and the 3rd International Intersex

Forum, all issued statements opposing nonconsensual genital normalization surgery and other human rights abuses faced by intersex people.[21] That same year, Australia passed an amendment that prohibits discrimination on the basis of intersex status, and Germany became the first country in Europe to allow parents of newborns without "clear gender-determining physical characteristics" not to register them as either male or female, but to choose a third blank box instead.[22] Also in 2013, the Heinrich Boll Foundation published a groundbreaking report by Dan Christian Ghattas, a representative of Organisation Intersex International.[23] Based on a survey of intersex individuals from twelve selected countries in the global south and east as well as Europe, Ghattas's report provides indisputable evidence of discrimination against intersex people and describes the needs of intersex activist organizations in these countries. Additionally, in 2013 a lawsuit was filed in South Carolina that challenged medically unnecessary sex-assignment surgery done on a sixteen-month-old child with an intersex condition.[24] This case—the first of its kind in the United States—was rejected by the Court of Appeals for the Fourth Circuit, but a state lawsuit in Richland County is currently moving forward.

In what follows, I do not explore these specific developments, but I cite them here to call attention to the uneven and escalating transnational momentum of intersex human rights activism. Instead, I analyze events that took place in an earlier phase of the intersex movement, the late 1990s, a period when the local and global trajectories of intersex activism were initially being charted. I focus on ISNA not only because it was the first intersex activist association in the world, but also because this organization was the first to realize the strategic potential of the discourse of human rights as a medium of political articulation for addressing intersex issues. My analysis does not imply that other intersex activist organizations have not been influential or that the examination of the strategies of those organizations, which both converge and diverge with those of ISNA, is any less important. Rather, analyzing ISNA's early activism through a transnational feminist lens offers valuable historical insight into the contemporary possibilities and limitations of grounding intersex advocacy in a human rights framework.

Transnational Feminist Perspectives on Human Rights

Celebratory accounts of human rights proliferate in the current era. However, framing human rights solely as a positive political achievement—as a protective shield from the intrusion of unjust exercises of power—risks obscuring the ways in which human rights also materialize unequal juridical, political, and socioeconomic relationships. Transnational feminist critiques thus draw attention to human rights as double edged. Recourse to human rights has undoubtedly enabled many oppressed groups (including women; sexual, racial, ethnic, and religious minorities; stateless people; and people with disabilities) to challenge exclusionary and unjust institutions and practices. Yet, although human rights instruments have empowered multiple marginalized constituencies around the globe, the moral universalism and US-/Eurocentrism of human rights tends to erase important differences among and between various political subjects.[25] By reducing power relations to a stark narrative of "us" versus "them," victims versus victimizers, human rights narratives are intrinsically exclusionary: they forget that different subjects arise through different histories and forms of power and that they may have different political desires and different conceptions of justice.[26] Moreover, such narratives oversimplify socioeconomic and political realities, conceal the complexities and enabling violations of subject formation, and obscure the interdependency of local and global flows.

From a transnational feminist perspective, three additional, interrelated issues need to be addressed to analyze the politics of contemporary deployments of human rights: 1) rights-based regulation, 2) neoliberalism, and 3) gender and sexual exceptionalism. I briefly consider each issue in turn.

1) *Rights-based regulation.* As legal and political technologies that guide the actions of individuals, institutions, and populaces, human rights are not merely empowering, but also regulatory. Wendy Brown explains this paradox in the context of feminist debates over women's rights as human rights as follows: "To have a right as a woman is not to be free of being designated and subordinated by gender. Rather, though it may entail some protection from the most immobilizing features of that designation, it reinscribes the designation as it protects us, and thus enables our further regulation through that designation."[27] Rights interpellate subjects within discursive and regulatory relations of power, knowledge, and normalization. Human rights must therefore be understood as not just offering positive juridical protection to otherwise vulnerable subjects, but as simultaneously producing and intensifying individuals' distribution across the "dispositif" of modern disciplinary subjectivity and biopower.[28] That is, to be a subject of human rights is to be subjectified *by* but also subjected *to* a range of distinct but often overlapping systems of governance. Many scholars have analyzed the ways in which communities and institutions have used sex/gender, race, class, religion, sexuality, ability, and other categories of difference to police the formal and informal parameters of citizenship and belonging.[29] These analyses demonstrate that national inclusion alone does not guarantee equal

rights or fair treatment. As Brown puts it, "rights almost always serve as a mitigation—but not a resolution—of subordinating powers."[30]

2) *Neoliberalism.* The mitigating function of rights is evident in a variety of global and local dynamics, especially neoliberal processes that fortify disparities between populations.[31] While its meaning is contested, for the purposes of this chapter I define neoliberalism in terms of the political economic principles that the international state system adopted during the last forty years. These principles include market deregulation, privatization, flexible accumulation, financialization, and personal responsibilization. The institution of these principles significantly restricts the social welfare function of the state while simultaneously promoting the interests of the "free market" and its attendant international division of labor in ways that reinforce racialized, gendered, and classed hierarchies.[32] Understood as such, neoliberalism has major implications for cultural politics. Some critics argue that neoliberalism undermines the promise of democratic freedom.[33] Others suggest that neoliberal projects of global restructuring both extend and rewrite older colonial legacies.[34] In these ways, economic globalization increases disparities between populations through the upward redistribution of wealth, by precaritizing lower- and middle-class communities, and by marking certain populations as disposable.

3) *Gender and sexual exceptionalism.* Human rights discourse not only is implicated in ongoing neoliberal transformations and the reordering of contemporary US and western empires, but also is linked with gender and sexual exceptionalism. US interests frequently position themselves as the authority on and exemplar of human rights, condemning human rights abuses in other countries while ignoring such abuses domestically. Iris Marion Young and other scholars and activists argue that the US government's use of Afghani and Iraqi women's human rights to justify the US-led "war on terror" fostered forms of gender exceptionalism that reified ideologies of American superiority and western universalism.[35] Jasbir K. Puar identifies sexual exceptionalism or "homonationalism" as an abiding feature of contemporary western LGBT rights movements.[36] Homonationalism is a form of LGBT patriotism that works with heteronormative citizenship by promoting marriage equality, compulsory monogamy, the nuclear family, consumerism, and militarism as the idealized expressions of national reproduction. Crucially, homonationalism also operates outside of the nation-state to elevate certain nations as more humane and liberal than others. In other words, gender and sexual exceptionalism not only naturalize and normalize but also attempt to globalize and falsely universalize contemporary US and western epistemologies and ontologies of sex/gender, sexuality, and citizenship. In so doing, these forms of exceptionalism

marginalize non-western concepts and practices of embodiment, subjectivity, and community.

By highlighting the issues of rights-based regulation, neoliberalism, and the imperial politics of US gender and sexual exceptionalism, transnational feminist perspectives reveal that human rights are not exterior to power, but are one of the mechanisms by which power—in its myriad forms—enables *and* constrains different subjects in dissimilar ways. Thus, deployments of human rights do not automatically manifest radical change. Rather, recourse to rights can challenge and disrupt, but can also reinforce, dominant ideological, political, and economic formations. This point is worth emphasizing because it focalizes the contextual nature of human rights; human rights advocacy is always shaped by a biopolitics and geopolitics of location. In the remainder of this chapter, I explore this thesis in relation to US intersex activism and its transnational travels.

The FGM Analogy

Though in its inaugural years ISNA often defined intersexuality through the language of queer (anti-identitarian) identity politics and sought to de-medicalize intersex altogether (as I explored in chapter 3), toward the turn of the millennium the organization undertook an effort to reach a broader audience, especially medical providers and parents of intersex children.[37] ISNA began to reframe intersex as a discrete set of embodied conditions, less an identity than an etiology. In embracing this view, ISNA implicitly reaffirmed the biomedical understanding of intersex as anatomically based even as it explicitly challenged the medical pathologization of people born with atypical sex characteristics.[38] It was during this period that ISNA discovered that the framework of human rights harbored useful resources for encouraging medical providers and the public at large to rethink the medical management of intersexuality. This adoption of human rights discourse by ISNA and other intersex activist groups reflected how various disenfranchised groups in social movements of the mid-twentieth century called for state-sanctioned legal and civil protections by appealing to internationally recognized instruments of human rights (including the Geneva Convention, the Nuremburg Code, the Universal Declaration of Human Rights, CEDAW, and the UN Convention on the Rights of the Child). This human rights framework, however, simultaneously generated important opportunities and created unique problems for intersex activists.

Framing intersex as a human rights issue, members of ISNA in the late 1990s began to compare nonconsensual intersex genital normalization surgeries with a fraught and contested practice of some African traditions referred to in mainstream western discourse as "female genital mutilation" (FGM). "To emphasize the likeness of intersex surgeries and female genital mutilation, ISNA's press releases in 1997 started referring to intersex surgeries as Intersex Genital Mutilation" or IGM.[39] Inserting their critique of western intersex surgery into highly charged debates about women's autonomy, cultural rights, and body politics within and across postcolonial and transnational contexts, ISNA activists narrativized intersex surgery as a violation of human rights similar to the violation that FGM is said to embody. This analogy thus sought to conjure and popularize a sense of moral outrage about intersex surgery that would mirror dominant global northern feminist and liberal humanist reactions to FGM.

Members of ISNA lobbied the US Congress in 1997 to include intersex as a protected category in a proposed federal statutory ban on "female genital mutilation." ISNA's argument was that American intersex surgery like FGM constitutes a violation of individual citizens' human rights to bodily integrity, informed consent, and personal autonomy. According to Chase, some anti-excision African migrant women in the United States were sympathetic to ISNA's argument, but white western feminists were reluctant to include intersex in their anti-FGM campaigns.[40] Noting the relative lack of western feminist and mainstream media attention to intersex surgeries in the context of high-profile debates about FGM and women's rights, Chase argues that the "othering" of African cultural practices by "first-world feminists" deflects attention away from genital cutting in the United States. That is, Chase critiques the complicity of American feminist gender essentialism with western imperialism, while also figuring intersex as a challenge to heteronormative and sexually dimorphic epistemologies (including feminist ones).

Chase extends this argument in a 2002 essay by arguing that the western understanding of clitoridectomy as "culturally remote"—always located "elsewhere," in distant geographic and temporal zones—"allows feminist outrage to be diverted into potentially colonialist meddling in the social affairs of others while hampering work for social justice at home."[41] Seemingly affirming postcolonial feminist critiques of western universalism, Chase argues that genital normalization surgery in the United States must be understood as a cultural—and not purely biomedical—practice. Chase's argument is based on analogics: likening western intersex surgery to FGM, Chase reads intersex surgery as a culturally specific manifestation of heteropatriarchal domination.[42] However, as this quote indicates, Chase also presumes that the organic unity of the nation-state prescribes or predetermines the appropriate contours and needs of "local" social justice agendas. The underlying implication of her critique of first-world feminist discourses on FGM is that "social justice at home" (in the United States), while connected with social justice struggles abroad, is nevertheless politically more important than, and ought to be privileged above, feminist concerns that arise in foreign affairs. Yet the very presumption that FGM is a "social affair of others," a "foreign" matter in the American context, is called into question by both the long history of the forced, coerced, and free migration of African cultural practices to the United States and by Congress's and US feminist activists' stated interest in banning a set of practices they uncritically deem "other" to the American way of life. From a transnational feminist perspective, neither "FGM" nor "IGM" can be said to discretely belong to a single nation-state or cultural tradition. One might even say that the intersex surgery/FGM analogy foregrounds the former to some degree at the expense of the latter. It is therefore deeply ironic that Chase critiques the "colonialist meddling in the social affairs of others" in this essay, as Chase was directing ISNA when the organization campaigned for the inclusion of "intersex" in the US federal ban on FGM.

By analogizing intersex surgery with FGM, Chase at once erases these complexities and reiterates rather than interrupts the "othering" of African cultural practices that she highlights. The intersex surgery/FGM analogy is especially problematic insofar as it consolidates the western/non-western and global north/global south binaries.[43] As is well known, during the 1980s and 1990s, Eurocentric discourses of so-called "global" or western liberal feminism denounced practices of FGM.[44] FGM was sensationalized in the western media, resulting in a sweeping criminalization of FGM in the United States and in many parts of the global north. Numerous western feminists who used FGM as a lever to call for the liberation of their "third-world sisters" relied on highly stereotypical, inaccurate, and frankly racist representations of "African" female genital cutting.[45] Such representations construct a monolithic account of FGM and African culture and subtend a series of troubling binaries that pit the "first world" against the "third world," the "west" against the "non-west," the global "north" against the global "south," and "enlightened" liberal feminism against "primitive" patriarchal societies.

These binaries obscure the fraught and complex histories of European empire, US settler colonialism, transnational capitalism, and modernity more broadly. They also reinscribe a patently neocolonial narrative that simultaneously secures and dissimulates the hegemony of American and European epistemic, cultural, and political economic formations. As Leslye Obiora argues, western anti-FGM campaigns cannot be adequately understood without critically attending to the roles that race and racism play in western feminist portrayals of African culture and African women as static and one-dimensional. Countering these stereotypes, Obiora points out that African practices of genital cutting are heterogeneous, as are the cultural milieus in which they occur. For this reason, she contends that mainstream western feminist discussions of FGM homogenize African culture and eclipse the diversity of African perspectives on genital cutting, extending and deepening troubling colonial legacies.

Wairimu Ngaruiya Njambi similarly critiques western anti-FGM discourse, noting that it erases the complex ways in which regionally specific cultural norms both enable and constrain the bodily agency of differently situated African women and girls.[46] "In presuming that bodies can be separated from their cultural contexts, the anti-FGM discourse," she writes, "not only replicates a nature/culture dualism that has been roundly questioned by feminists in science studies and cultural studies, but has also perpetuated a colonialist assumption by universalizing a particular Western image of a 'normal' body and sexuality in its quest to liberate women and girls."[47] In an autoethnographic analysis of her own experience as a circumcised Gikuyu woman, Njambi argues that female circumcision does not have a univalent, monolithic, or self-evident meaning, and cannot be adequately understood outside the cultural contexts in which it is practiced. Njambi observes that within the gender order of Gikuyu culture, young women of her social group largely viewed circumcision as a "rite of passage," and that the surgery earned them a newfound sense of adulthood, confidence, and respect from others. "By completing my irua, I became a Gikuyu woman . . . I entered a category whose pleasures and benefits had previously been denied to me."[48] Acknowledging that some African feminists disagree with her interpretation of the enabling possibilities of circumcision, Njambi writes: "Hopefully, what my story conveys is that there are ways of looking at the female circumcision issue which go beyond colonialist stories of barbarity and primitivity; stories that surely leave the represented without a sense of agency. Practices of female circumcision involve negotiations, ambiguities, complexities, and contradictions that must be addressed and not dismissed, even as we problematize them."[49]

By addressing the complex forms of agency and power exercised through African forms of genital cutting, Njambi's analysis offers an alternative to the reductionism of western feminist anti-FGM discourse. Her work reveals that creating more effective cross-border dialogues about body politics requires a heightened sensitivity to the ways different subjects arise through different histories and forms of power and therefore may have different conceptions of the ethics of embodiment, agency, and justice.[50] As Claire Hemmings explains her reading of the debates surrounding

Njambi's intervention: "what cannot be contemplated in . . . Western feminist accounts . . . is that the practice of FGC may be experienced neither as an oppression to be left behind nor an unspeakable horror or even unpleasant necessity. What is inconceivable is that any FGC practice may be actively embraced, pleasurably anticipated, and experienced as a marker of becoming an adult or becoming a member of a desired community."[51]

The moralism of western feminist debates on FGM not only silences the wide range of African interpretations of genital cutting, but also prevents western feminists from addressing African activists as legitimate political agents. Deflecting attention from questions of African women's general well-being, it forecloses opportunities to construct more genuinely transnational feminist alliances.[52] From this perspective, one might say that the intersex surgery/FGM analogy instantaneously discloses and effaces the asymmetries of contemporary transnational power relations. Put differently, ISNA's deployment of the analogy exposes the organization's failure to reflexively account for its own location of power in the global north in relation to African women specifically and the global south more generally.[53]

The intersex surgery/FGM analogy also obscures crucial differences between these disparate practices and their regulation in western societies.[54] Intersex surgeries seek to reduce any sexual "ambiguity" at infancy in order to promote sexual dimorphism and heteronormative gender development and are sanctioned and performed by western-trained biomedical experts. Genital cutting in Africa, by contrast, tends to occur in developing, postcolonial rural communities that lack access to potable water and other resources and are performed by local cultural experts whose primary concern is not the regulation of sexual ambiguity but rather the maximization of a certain cultural ideal of feminine corporeality frequently tied to the chastity, modesty, and marriageability of young women.[55] While such examples raise important questions about the heteropatriarchal regulation of African women's bodies in some tribal contexts, as Njambi argues, circumcision may also be understood as a bodily transition into a new form of subjectivity affording its own distinct pleasures, agentive capacities, and ethical responsibilities.[56]

Consequently, the intersex surgery/FGM analogy raises both strategic and ethical questions about intersex politics. According to Ben-Asher,

> These two types of genital surgeries are exceptionally different, in time, place, and ideology, and their merger in legal strategy erases these crucial differences. Furthermore, there is an ethical concern when group projects of de-subjugation undercut each other. Thus, intersex politics that is insensitive to the Western

> normalization of non-Western African traditions exchanges one social harm for another.[57]

Underscoring the regulatory operations of different forms of power, Ben Asher recognizes that the politics of FGC and intersex surgery are qualitatively and quantitatively different "in time, place, and ideology." By piggybacking onto anti-FGM campaigns, ISNA activists oversimplified these differences, a gesture that in turn obscures the "Western normalization of non-Western African traditions." In effect, the analogy presumes that the appropriate response to intersex surgery and FGM must be one and the same: the imposition of US neoliberal democratic frames of subjectivity. The analogy posits the neoliberal "autonomous" individual as the telos of both intersex and anti-FGM activisms. In this way, it perpetuates a US exceptionalist and westocentric understanding of human rights.

ISNA's lobbying effort to include intersex surgery within the legislative ban on FGM ultimately failed. Specifically, activists were unable to convince the US Congress that intersex surgeries fall under the category of "cultural practices" and are therefore not "medically necessary." The legislative ban explicitly makes an exception for cases of "medical necessity," stating in the pertinent section that "a surgical operation is not a violation . . . if the operation is . . . necessary to the health of the person on whom it is performed and is performed by a person licensed in the place of its performance as a medical practitioner."[58] Using biomedical authority to legislate the boundaries of health, the ban presumes that the "health of the person" is objectively measurable, when in fact historians of medicine have shown that "health" is a variable cultural construction.[59] As Njambi points out, many legal and fashionable forms of body modification in the global north—such as tattooing, piercing, penis/clitoris slicing, tongue slicing, and cosmetic procedures, including Botox injections, liposuction, breast and posterior implants and augmentation, and vaginal rejuvenation—might be viewed as potentially unhealthy, especially if we recognize that their performances are sometimes compelled by powerful regulatory ideals.[60] The factors that determine whether a procedure, bodily regimen, or corporeal way of life is considered healthy are themselves culturally, politically, and economically overdetermined.

Instituting a binary opposition between health and culture, the statute bans genital surgeries that are performed for cultural purposes (such as the transmission of traditions, rituals, or values), which it views as inherently unhealthy. That is, the statute maintains a more general distinction between culture and science that places intersex under the jurisdiction of biomedical expertise. The ban thus criminalizes FGM on cultural grounds while it legitimates intersex surgery on medical grounds. Crucially, the ban simultaneously others African bodies and American intersex bodies, but it does so in distinct ways. Taken together, these othering practices underwrite the ban's US-centric, neoimperial understanding of what constitutes a normal and healthy body.

The Colombian Constitutional Court

While ISNA was unsuccessful in its bid to convince the US Congress that intersex surgery is medically unnecessary along the same lines as FGM, ISNA in the late 1990s was able to exert its influence in other transnational arenas. In 1999, ISNA redeployed the intersex surgery/FGM analogy in an amicus brief it submitted to the Constitutional Court of Colombia that played a crucial role in two decisions by the court, which proposed a new standard of informed consent and banned parents from giving consent to normalizing genital surgeries for minors aged five and older. On a comparative global scale, these decisions were historic. Colombia stands alone as the only country in the world that provides legal protections against the nonconsensual medical normalization of intersex individuals.

ISNA issued a press release on the occasion of the Colombia decisions that reads as follows: "Colombia High Court Restricts Surgery on Intersex Children."[61] ISNA's characterization of the 1999 decisions as a "restriction" is somewhat misleading, however, as the Colombian Constitutional Court's 1999 rulings actually revised an earlier 1995 ruling by the court that banned normalizing infant genital surgeries across the board. In 1995, a young male-identified petitioner whose penis was accidentally ablated during what was presumably a rudimentary circumcision and who was surgically reassigned as female brought his case before the Colombian Constitutional Court. The petitioner, who said he never fully developed a female "gender identity," argued that the surgical sex assignment violated his human rights to self-determination and autonomy. The court ruled in favor of the plaintiff, finding that parents do not have the right to consent to cosmetic genital surgeries on a child. Summarizing this case, ISNA activists termed it "Colombia's own 'John/Joan' case," referring to the much-publicized story of David Reimer.[62] As this analogy reveals, ISNA viewed the 1995 Colombia decision through a US-centric lens that posited the North American experience of genital surgery and its intersex activist critique as the authoritative standard by which to comprehend intersex issues internationally. The comparison is problematic because it obscures the particularities of Colombian citizens' distinct experiences and analyses of anatomical sex variation and genital surgery, and also because it figures the United States as both the origin and defining horizon of intersex political advocacy. I am not suggesting that such comparisons cannot be drawn, but that any effort to do so must account for the uneven locations of power, epistemic authority, and discursive agency occupied by US intersex activists vis-à-vis their Latin American counterparts.

In the wake of the 1995 ruling, Colombian surgeons specializing in intersex treatment found themselves in a double bind. Their medical training and the law were at odds. Paradoxically, they continued to recommend normalizing surgeries to parents of intersex children, but they refused to perform the surgeries in an effort to stave off the threat of legal retribution. This situation came to a head when, in 1999, two lawsuits were filed. In each case, the parents of a child diagnosed with an intersex condition asked the Constitutional Court of Colombia to approve normalizing genital surgeries. Ben-Asher summarizes the findings of the court as follows:

> The court once again invalidated parental consent, recognizing that: 1) intersexed people may constitute a minority entitled to protection by the state against discrimination; 2) "corrective" surgery may be a violation of autonomy and bodily integrity motivated by the intolerance of parents toward their children's anatomy; 3) parents are likely to make decisions based upon their own fears and concerns rather than what is best for the child, especially if they are pressed to decide quickly; 4) a new standard of consent, "qualified, persistent informed consent," must be adopted in order to force parental decisions to take into account only the child's interest; and 5) for children over five years old, parents cannot consent, because the child has achieved an "autonomy" that must be protected, and because the child has already developed a gender identity. Thus, the consent of the parents of an eight-year old was invalidated by the court because the child was too old for a surrogate to consent on her behalf.[63]

To arrive at its decisions, the court conducted an extensive survey of medical, psychological, ethical, and legal experts in Colombia and abroad.[64] The decisions reference the Convention on the Rights of the Child, several opinions by the European Court on Human Rights, and previous opinions of the Constitutional Court of Colombia on protections for homosexuals and transsexuals.[65] In addition, the court cites the amicus brief submitted to the court by ISNA. By referencing ISNA as a legitimate authority on intersex issues, the court gave important international validation to ISNA's activism. However, the court did not follow all of ISNA's suggestions. The court's partial adoption of ISNA's recommendations can be read as part of a larger and longer process of the transnational circulation and solidification of US and western understandings of intersex, embodiment, subjectivity, and human rights in the global south, even as these understandings have been transformed and challenged by local actors.

To clarify what I mean by this claim, first it is important to situate the court (Colombia's highest judiciary) and the politics of sex/gender and intersex in Latin America within a wider historical context. A former Spanish colony, Colombia is the oldest constitutional government in Latin America. Democratic liberalism as formal political doctrine has existed in Colombia since the country won independence from Spanish rule in the nineteenth century. Since the early twentieth century—when President Theodore Roosevelt executed "gunboat diplomacy" to back the separatist revolution in Panama so as to secure US control over the construction of the Panama Canal—US influence in the region has been substantial. Colombia has been a key site, among others in Latin America, for the making of US empire and the development of neoliberal economic policy.[66]

These historical processes have shaped the politics of sex/gender in Colombia and Latin America more broadly in particular ways. According to Vek Lewis, "sexual diversity and sex and gender variance exhibit differences in Latin American locales and produce identity logics and formations distinct from those found, for instance, in the Anglophone World." However, homosexualities and trans and intersex identities also exist in Latin America. "Some of these are not dissimilar to those of the Anglophone world."[67] Such divergences and convergences are increasing in an age characterized by globalization. To critically understand historical discourses about sex/gender variance in Latin America, Lewis argues, it is important to interrogate the colonial legacies that have shaped these discourses. As numerous scholars have demonstrated, Spanish and other western imperialists pathologized and marginalized pre-Colombian sex/gender systems during efforts to impose colonial-modern conceptions of sex/gender and sexuality across the region.[68] Attending to such legacies, I would suggest that the twentieth-century medicalization of people with unusual sex anatomies in Colombia and some other Latin American nations can be read as a manifestation of what Anibal Quijano calls the coloniality of power: the ways in which colonial epistemologies and ideologies of sex/gender, sexuality, race, and class have outlived formal territorial colonialism and persist in contemporary societies.[69]

Intersex politics—both in the United States and globally—cannot be disaggregated from genealogies of imperialism, settler colonialism, and racial formation that have fundamentally shaped longstanding Euro-American eugenic, Malthusian, and sexological ideas about the relative degree of sexual development of Anglophone versus non-Anglophone racial groups.[70] While Quijano's thesis remains underexplored in the field of intersex studies, a few scholars have pursued notable work in this direction. Hilary Malatino observes that colonization was central (both literally and metaphorically) to the consolidation of scientific disciplines—including teratology, eugenics, and sexology—that study the appearance and development of monstrous bodies.[71] The medicalization of intersexuality can be seen as an effect of western scientific efforts to rationalize fleshly bodies into the normative constraints of the colonial-modern social order. Zine Magubane suggests that race, imperial history, and national context have played vital but underanalyzed roles in the production and reproduction of the concept of intersex.[72] And María Lugones argues that the "sexual fears of colonizers led them to imagine the indigenous people of the Americas as hermaphrodites or intersexed, with large penises and breasts with flowing milk. But as Paula Gunn Allen and others have made clear, intersexed individuals were recognized in many tribal societies prior to colonization without assimilation to the sexual binary."[73] Lugones contends that colonialism established a new sex/gender system that created very different arrangements for colonized subjects than for white bourgeois colonizers. "Not all different traditions," Lugones writes, "correct and normalize intersex people. So, as with other assumptions, it is important to ask how sexual dimorphism served and continues to serve global, Eurocentered, capitalist domination/exploitation."[74]

These scholars reframe the global history and politics of intersex in important ways. They reveal that the medical normalization of intersexuality cannot be reduced to narratives that focus solely on the "stigma and trauma" caused by human rights violations, as ISNA argued.[75] Nor is medicalization only about the regulatory production of sexual dimorphism and binary gender. Rather, debates about whether and how to respond to bodies in doubt are also overdetermined by national context and classed and racialized logics. From a transnational feminist perspective, then, the western medicalization of people with nonstandard sex anatomies can be interpreted as a key element of biopolitics: "an explosion of numerous and diverse techniques for achieving the subjugations of bodies and the control of populations."[76]

This background makes it possible to further contextualize the various transnational knowledge flows that are entangled in the court's decisions. The court draws on liberal political philosophy, ISNA's human rights approach to intersex, and western biomedical understandings of the body. And it is at the intersection of these three elements that the racialized resonances of the court's decisions begin to emerge. In particular, the court follows ISNA's proposition that it is important to question the medical *necessity* of intersex genital surgery. The court's decision states that "the actual treatments have caused harm, and there is no convincing evidence that they are necessary or beneficial."[77] Supporting this claim, the court cites ISNA's amicus brief, which invokes "both the Nuremburg Code and the basic principles of human rights law," which "prohibit subjecting a child to involuntary, irreversible, and medically unnecessary genital surgeries."[78] Crucially, in this section of its brief ISNA redeploys the FGM analogy in its bid to influence the court:

> It is repugnant and contrary to a child's basic human rights to allow a parent to consent to medically unnecessary genital surgery for the purpose of dictating the child's future gender identity or of altering the child's body to conform to an idealized cultural notion of "normal" genital appearance. This principle has been established in the analogous context of female genital mutilation, where a wide variety of human rights authorities and organizations have determined that involuntary genital surgery performed on female children violates basic human rights to bodily integrity and personal dignity and autonomy . . .
>
> Many human rights bodies have condemned female genital mutilation, defined as the removal of all or part of the clitoris, inner labia, or outer labia. "Feminizing genital surgery" reduces the size of the clitoris by removing parts of the clitoris . . . Clitoral reduction surgery is thus clearly covered by the definition of female genital mutilation.

By the end of this passage, ISNA transforms the analogy into an equation, positing intersex surgery as a form of FGM. Doing so, ISNA frames intersex surgery as a retrograde cultural practice that is not in keeping with western society's commitment to "bodily integrity and personal dignity and autonomy." ISNA treats these ideas of bodily integrity, personal dignity, and autonomy as irrefutable universals, but they are in fact culturally specific. Moreover, when deployed as universals, these ideas lend themselves to western progress narratives that presume the global superiority of Euro-American formations of modernity, civilization, and development, narratives whose racialized logics position the global south and its inhabitants as perpetually underdeveloped. By proposing that the Colombian Constitutional Court understand intersex surgery through the lens of western human rights generally and white, liberal US feminist representations of FGM specifically, ISNA not only discursively reiterated once more the "Western normalization of non-Western African traditions."[79] This time, it also attempted to export the US-centric racialized epistemology that naturalizes such processes of normalization into the letter of the Colombian law.

This gesture is especially confounding because in the late 1990s, anti-FGM activism did not have a significant history or presence in Colombia.[80] Perhaps for this reason, the court's decisions do not explicitly reference ISNA's equation of intersex surgery with FGM. However, the court does use the Spanish word "mutilación" (which translates into English as "mutilation, defacement, or emasculation") to describe intersex genital surgery numerous times in its decisions.[81] The court's invocation of "mutilación" reflects ISNA's thesis that intersex surgery is not only medically unwarranted but also potentially harmful. Additionally, rendering intersex surgery as "mutilación" inscribes a negative moral value on the practice. It folds intersex surgery into the moral calculus of hegemonic western discourses of civilization and modernity. These discourses regard certain practices of body modification or comportment that are coded as largely western/civilized/white (infant male circumcision, or cosmetic surgery, for instance) as intelligible, valuable, and acceptable, and various others that are coded as non-western/uncivilized/nonwhite (such as FGM; ear, lip, and neck stretching; and veiling) as unintelligible, improper, and unacceptable on the basis of highly racialized and gendered US- and Eurocentric regulatory ideals. The court's usage of "mutilación" reproduces the moral universalism and cultural logics at work in the transnational circulation of ISNA's FGM analogy.

A close examination of the relationship between ISNA's amicus brief and the court's decisions reveals that western understandings of intersex, embodiment, subjectivity, and rights travel across national borders; and that, in the process, local actors rearticulate these understandings in both familiar—that is, the court's deployment of "mutilación"—and sometimes new and surprising ways. With regard to the latter, in an important development, the court suggests that the standard category of informed consent in liberal doctrine is not sufficient to protect intersex minors older than five years of age, and that a new category is needed. The court names that new category "informed, qualified and persistent consent."[82] According to the court, should the minor in this case seek to authorize a medical intervention, "her consent needs to be qualified," by which the court means that her decision needs to be accompanied by treatment and counseling by an "interdisciplinary team" who will provide "not only psychological support but also develop a method to ensure that the patient's authorization is informed and genuine."[83] As the court's language implies, the category of "informed, qualified and persistent consent" is designed to enhance the autonomy of intersex minors even as it still subjects them to a different (and supposedly more beneficent) form of medicalization.

Some observers have critiqued the court's decisions on precisely this point. Patricia González Sánchez, Catalina Velásquez Acevedo, and Sandra Patricia Duque Quintero argue that the court's new standard of informed consent provides a useful legal tool for protecting intersex minors five years of age and older from nonconsensual medical normalization.[84] However, they suggest that the court's decisions may nevertheless contravene the 1991 Colombian Constitution's commitment to pluralism, in that the court maintains the provision that intersex minors must be assigned to one of two sex categories. Reading sexual dimorphism not as a natural fact but as a socially imposed norm, Sánchez, Acevedo, and Quintero contend that enforced sex assignment exposes tensions between the principle of beneficence and the principle of autonomy. Additionally, they point out that the court's new standard of informed consent does not effectively address the wider problem of the stigmatization of intersexuality in Colombian society and that its decisions reinforce rather than challenge medical paternalism.

Morgan Holmes makes a related argument in her analysis of the case in "Deciding the Fate or Protecting a Developing Autonomy? Intersex Children and the Colombian Constitutional Court."[85] She notes that "the court's decision is not made independently from the information it demands of the medical community, and its decision does not supersede the authority of medicine to define what counts as sexual anomaly."[86] Holmes continues:

> The decision does not undermine the authority of medical knowledge or of practitioners to explain to parents and families what the proper course of action should be when an intersexed child *is born*. In its worst potential implications and uses, the court's decision may simply amplify the need to expedite procedures, making sure they take place in the neonatal period before the infant has acquired any self-awareness at all.[87]

While the court imposes restrictions on genital surgery for children older than five years of age and offers an expanded definition of informed consent, the court still leaves it to biomedical experts to define intersexuality as a set of conditions that require clinical treatment. Furthermore, the court continues to privilege parental rights above children's rights. For these reasons, Holmes suggests that ISNA's enthusiastic endorsement of the court's decision as a victory for intersex people's human rights "neglect[s] the more subtle maneuvering of the court on this point." By giving juridical legitimacy to infant genital surgeries performed under certain conditions, the court explicitly ignores "every human being's right to bodily integrity."[88] That the court fails to recognize this right "indicates that the court is not actually as interested in protecting children's autonomy as it first appears."[89]

In her 2011 article "Estados Intersexuales en Menores de Edad: Los Principios de Autonomía y Beneficencia," Julia Sandra Bernal Crespo suggests that the court fails to ensure autonomy because it does not empower children with the right to *not* identify as either male or female.[90] Crespo argues that Colombian citizens should have a right to equality without any requirement of the "anulación de la diferencia," or the "cancellation of difference."[91] Criticizing the court's reliance on a pathologizing colonial-modern view of intersexuality, Crespo claims that intersex states should not be considered disorders, but rather variations of sexual differentiation.[92] Crespo suggests that materializing intersex rights within the context of neoliberal restructuring in contemporary Colombia requires both a scientific and a cultural paradigm shift. The scientific paradigm shift must be depathologizing, while the cultural paradigm shift would need to focus on revaluing bodily diversity. Such paradigm shifts, Crespo concludes, would provide Colombian law with a stronger basis for insuring the autonomy of individuals to determine their own corporeality and identity.

These critiques of the Colombian Constitutional Court's decisions demonstrate that the meaning of intersexuality continues to be contested within and beyond the boundaries of the Colombian nation-state. While the court's decisions remain historically unprecedented and clearly highlight the global impact of US intersex activism, they also more subtly reveal the biopolitical collusions that are reticulated in the transnational regulation of sexed bodies. These collusions (between the Colombian legal system, western biomedicine, the coloniality of power, neoliberal technologies of subject regulation, and US intersex activism) are tightly wound. I have attempted to begin unraveling them here by analyzing the complex relationships between variously sexed and gendered subjects in diverse locations. The mediations, betrayals, and alliances encoded within those relationships are multiple and fraught. Nonetheless, critically decoding them makes it possible to begin imagining intersex citizenship otherwise.

Conclusion

Both the Colombian Constitutional Court's decisions and the US legislative FGM ban rely heavily on western biomedicine as a source of knowledge about the meaning of and proper way to respond to and treat intersex bodies. The project of provincializing the intersex imaginary—of linking ways of thinking about intersex bodies to specific places and histories—reveals that these strategies depend on particular assumptions about embodiment, medical beneficence, and subjectivity that deserve interrogation and critique. ISNA's FGM analogy and the organization's influence on the Colombian Constitutional Court thus raise the problem of "intersex imperialism"—namely, the imposition of western conceptions of atypically sexed bodies and of how best to treat and respond to such bodies.[93] Decolonizing intersex requires thinking about atypically sexed bodies otherwise—in a different language, so to speak. As the Argentinian philosopher and intersex and trans activist Mauro Cabral puts it, the problem of intersex imperialism raises

> three considerations about language: (1) for those who, like me, speak, write, breathe, love and fight in a language other than English, decolonizing the way in which we are named multiplies to include not only medical speech but also the terms imposed by the Anglophone hegemony within international conversations on intersex; (2) the construction of an international intersex movement not only has depended—and still depends—on that hegemony, but has also considerably restricted the possibilities of other ways of communicating, including those of poetry, fiction, erotic, and other manifestations of "inefficient" speech; (3) even against this double suture, the word "intersex" still designates a persistent question, a scarred question that lies from tongue to language, a scar never fully healed, constantly reopened.[94]

As Cabral emphasizes, decolonizing intersex requires taking seriously the ethical imperative to learn from and form solidarities with indigenous and emergent traditions that offer alternative languages, cartographies, and valuations of bodily difference, while also guarding against the dangers of cultural appropriation. It requires animating oppositional and differential modes of consciousness that deconstruct the presumed superiority of western knowledges of the body and frames of subjectivity.[95] In other words, decolonizing intersex necessitates a broader recognition that "the word 'intersex' still designates a persistent question, a scarred question that lies from tongue to language, a scar never fully healed, constantly reopened." Coming to terms with that constantly reopened, never fully healed scar/question entails learning to bear witness to a trauma that ceaselessly reproduces itself, marking intersex bodies with the scalpel and the suture, rendering them as the exceptions that give the lie to the normalizing corporeal rule. Decolonizing intersex, then, also entails fostering an emphatically unnatural tolerance for a pain at once individual and collective, which also paradoxically unfolds as a holistic capacity for epistemological disobedience and an ethical willingness to be undone and remade by our relationships with others.

As preparatory work for that project yet to come, this chapter has attempted to provincialize US debates about intersex—that is, to explore how US ideas about human sex variation, human rights, and personhood that appear universal are "also, at one and the same time, drawn from very particular intellectual and historical traditions that [cannot] claim any universal validity."[96] Movements for intersex human rights in the global north and west that do not self-reflexively question their overdetermination by colonial legacies, biopolitics, and geopolitical power disparities not only risk erasing and othering the distinct histories, epistemologies, subjectivities, sex/gender systems, and bodies of peoples in the global south and east; they also undercut the formation of cross-border alliances to advance struggles for sexual and gender justice. The challenge is to create multilingual and multicultural dialogues about body politics that can acknowledge the messy, precarious, and asymmetrical relationships between differently sexed and gendered subjects in diverse locations. Transnational feminist perspectives can help to challenge intersex imperialism by emphasizing that critical scrutiny of the politics of location of *all* intersex activist projects is essential for building more ethically accountable transnational movements for corporeal freedom.

5

Intersectionality and Intersex in Transnational Times

This chapter builds on the previous ones by asking: what is the relationship between the paradigm of intersectionality and debates about intersex in transnational times? Sally Markowitz's 2001 article "Pelvic Politics: Sexual Dimorphism and Racial Difference" provides a useful starting point for this chapter's exploration of this question.[1] I begin by engaging Markowitz in some detail because, on the one hand, she offers a highly innovative but under-cited genealogy of racial formation as central to the meaning and materiality of sex/gender in late modernity, and, on the other hand, because her work also raises the specter of the under-interrogated place of intersex, trans, and gender nonconforming subjects within that genealogy. I want to emphasize at the outset that I have learned much from Markowitz's work. In questioning Markowitz's unmarked bifurcation of intersectionality from intersex, a gesture that privileges the former to the exclusion of the latter, this chapter seeks to show that the deconstruction of this bifurcation can provide stronger resources for writing more inclusive and capacious genealogies of racialized gender that incorporate intersex, trans, and gender nonconforming subjects—a project that would not be possible without Markowitz's groundbreaking intervention.

In "Pelvic Politics," Markowitz investigates the role of the pelvis in ideologies of sex and race. Examining the history of evolutionary biology, specifically the work of turn-of-the-century British researcher and physician Havelock Ellis,[2] Markowitz argues that the very idea of sexual difference emerged through the racialization of bodies enacted by European imperialism and its authorizing scientific projects. Ellis was a founding father of sexology, a proponent of eugenics, and a major influence on the intellectual development of Sigmund Freud, Alfred Adler, Alfred Kinsey, and, among many others, John Money.[3] Analyzing the racialized resonances of Ellis's research on sexual psychology, Markowitz contends that "the ideology of sex/gender difference itself turns out to rest not on a simple binary opposition between male and female but rather on a scale of racially coded degrees of sex/gender difference culminating in the manly European man and the feminine European woman."[4] Despite her ahistorical use of the category gender here and throughout her article,[5] Markowitz's larger point is at once theoretically arresting and politically illuminating: if the ideology of sexual dimorphism is always already constituted in and through processes of racial formation, then sexual difference cannot be as binary as it might initially appear to be.

As Markowitz documents, so-called male and female traits are thoroughly racialized in the archives of the eugenic sciences. These racialized interpellations of sexual difference have also become deeply naturalized in western cultural formations. "In dominant Western ideology a strong sex/gender dimorphism often serves as a human ideal against which different races may be measured and all but white Europeans found wanting."[6] Sexologists like Ellis claimed that humanity was composed of a "great chain of being," a multilayered hierarchical structure that superordinated the "manly European male" and his heteronormative counterpart, the "feminine European female" as the idealized subjects of natural sexual dimorphism.[7] This eugenic hierarchy simultaneously marked non-European males and females (especially those of African, Asian, and indigenous American descent) as subordinate, or less "sexually developed" in evolutionary terms. For Ellis, not all female bodies could be labeled "feminine," because he assumed that femininity achieved its developmental telos exclusively in the biology of the Anglo-European female. For this reason, Markowitz suggests that sex and race are not actually distinct categories but are rather co-constitutive and co-dependent. They are, as many have argued, intersectional.

As a keyword in contemporary theory and culture, intersectionality is usually defined as the study of interesting systems of racial, sexual, and class stratification; discrimination; and domination. Hailed as one of the most important and transformative contributions of women of color scholarship to women's, gender, and sexuality studies, and to interdisciplinary research more broadly,[8] intersectionality encompasses a diverse range of approaches that a) reject the idea that subjectivity and sociality are formed by any single axis of oppression (such as race *or* gender); and b) instead frame subjectivity and sociality as inherently complex, multiply overdetermined, and formed by multiple interlocking systems, relations of power, and categories of difference that include race, gender, *and* class, among others. Coined in 1988 by Kimberle Crenshaw and developed primarily in scholarship and activism by US women of color, particularly black feminist theorists and organizers,[9] intersectional frameworks hold that oppressions do not act independently of one another but rather intersect to shape people's life chances in various systematic and unequal ways, creating multilayered systems that privilege white, middle-class, heteronormative folks while simultaneously subordinating and women of color, queers of color, and other historically disenfranchised groups.

Markowitz suggests that feminists have both overlooked and struggled to come to terms with sexual dimorphism as a racialized concept for several reasons: the "whiteness" of mainstream feminism, the disciplinary separation of race and sex/gender as analytic categories and objects of knowledge in academe, and the tendency within feminist theory to emphasize binary opposition as the primary modality of power.[10] According to Markowitz, this situation has led some contemporary feminist and queer scholars to overemphasize the need to think "beyond sexual dimorphism," as the subtitle of one widely cited anthology puts it,[11] as the imperative next step in the project of denaturalizing normative ideas about sex/gender.[12] Markowitz rebukes this project because, in her words, it "misses a very important point: An ideology that considers sexual dimorphism to be embodied only in European 'races' has *already*, in a sense, thought beyond it—hardly, it starts to seem, a revolutionary accomplishment."[13] With this pointed critical characterization of feminist, queer, trans, and intersex critiques of sexual dimorphism as thoroughly un-revolutionary and complicit with the perpetuation of white supremacy and its colonial legacies, Markowitz challenges scholars and activists to more carefully attend to how sex and gender are racialized and classed in genealogies of biopolitics and geopolitics.

Although Markowitz does not discuss the history of intersex in sexology and eugenics in her article, she does briefly mention the topic. Markowitz writes: " 'Thinking beyond sexual dimorphism' is far from a nearly impossible, utopian feat, requiring studies of non-Western cultures or physical anomalies even to give us the idea; on the contrary, to grasp a *racialized* binary of sex/gender is already, in a sense, to have thought beyond what now appears sex/gender ideology's rather superficial binarism."[14] Markowitz's use of the adverb *already* implies that thinking sex/gender as a racialized concept renders sex/gender ideology's binarism "superficial" and obviates *in advance* questions about the place (or nonplace, as it were) of bodies regarded as atypical in processes of racial/sexual formation. Markowitz's rhetoric makes it seem as if analyses of intersex, trans, and gender noncomformity in both western and non-western contexts do not have anything of significance to add to efforts to forge more nuanced forms of antiracist feminist inquiry. It is this particular way of framing studies of racialized sex/gender as superordinate to studies of corporeal variation (as if the former do not always *already* intersect with the latter and vice versa) that I want to interrogate and challenge in this chapter. Thus, my aim is to refuse the bifurcation of intersectionality from intersex by reading both against their ingrained assumptions and through their intellectual and political rifts.

From a critical intersex studies perspective, Markowitz's argument is troubling because it reflects a more general trend within institutionalized US discourses of intersectionality. What are the implications of labeling sex/gender binarism as "superficial?" Does this gesture not obscure the ways in which binarism has been variously rearticulated, reinforced, and normalized in different ways within and across settler colonial, neocolonial, postcolonial, and transnational contexts? Rather than dismissing "studies of non-Western cultures or physical anomalies [sic]," the project of articulating more effective forms of anti-racist feminist and queer critique would be better served, I contend, by attending to how sex/gender binarism is both reproduced and called into question through uneven and dissimilar circuits of racial, sexual, and national formation that are also often paradoxically interconnected.

By and large, scholars of intersectionality have not attended to intersex matters, nor have they explored how genealogies of intersex might speak to, challenge, or reconfigure intersectionality as an analytic paradigm. In a strikingly similar way, critical intersex scholarship has infrequently analyzed the implications of intersectional analytics for rethinking intersex embodiments and identifications.[15] This chapter seeks to address both lacunas. In my view, it is not enough to say that unmarked references to sex/gender reify the presumed whiteness of that category, are falsely universalizing, and then to leave the analysis stalled there. Rather, we need to develop critical accounts of racial/sexual formation that center women and queers of color as well as intersex, trans, and gender nonconforming folks. This chapter therefore asks: what is the relationship between intersex and intersectionality in a globalizing world?

Bringing together intersex studies with intersectional and transnational queer feminist perspectives, this chapter considers the controversy and debates surrounding South African professional middle-distance runner Caster Semenya. The Semenya controversy offers an opportunity to inquire into how western biomedicine, transnational sporting cultures, and ideologies of sex, gender, race, and nation encode representations of extraordinary bodies, simultaneously reproducing and contesting sex/gender binarism in different ways in and across different national contexts. In the first section of this chapter, I closely read western media representations of the Semenya controversy. In the second section, I investigate how athletic officials, activists, and scholars problematically linked Semenya's treatment to intersex issues, despite the fact that Semenya is a woman and has never publicly identified as intersex. In the third section, I analyze responses by South African fans, athletic officials, and politicians, which highlighted the complex entanglement of anticolonial resistance and Black Nationalism with the reproduction of heteropatriarchal sex/gender binarism in South Africa. In the fourth section, I explore the racial and national overdetermination of bodies in doubt to understand why the Semenya controversy became a lightning rod for competing intellectual claims about intersex rights, on the one hand, and intersectionality and racialized gender, on the other. Throughout this chapter, I build on Jennifer Doyle's work in an effort to further elaborate on how "the incommensurate systems of meaning that shape Semenya's story"[16] converge and diverge, exposing key tensions between the different narratives, theories, and political desires that were imposed on Semenya to make sense of her story.

Semenya's story, it turns out, was never entirely her *own* story at all. Rather, Semenya's story became a history of the appropriation of her story for the agendas of western and non-western elites. At this chapter's end, I therefore consider how Semenya herself spoke back to these impositions and appropriations—through her own statements as well as her athletic accomplishments—and thereby implicated athletic officials and fellow competitors, fans, politicians, scholars, and activists in the silencing of subaltern speech.[17]

The Biopolitics of International Sex Testing

After winning the gold medal in the women's 800-meter dash at the International Association of Athletics Federations (IAAF) World Championship in track and field in Berlin on August 19, 2009, eighteen-year-old South African middle-distance runner Caster Semenya quickly became the subject of intense western media scrutiny. In a *New York Times* article published the day following her win titled "Gender Test after a Gold-Medal Finish," sports journalist Christopher Clarey reported that, in the wake of Semenya's win by a margin of more than two seconds (still two seconds below the world record), the IAAF, under pressure from fellow competitors, required Semenya to undergo a barrage of "sex-determination testing to confirm her eligibility to race as a woman," which some experts quoted in the *Times* also called a "gender test."[18]

As Doyle observes in her analysis of media representations of Semenya, "she was accused by many of being a man, of being not '100 percent' woman. She was diagnosed in headlines as a hermaphrodite, as intersex, as a gender freak."[19] As Judith Butler has argued, to have one's status as a coherently sexed/gendered subject called into question is often a violent and violating process of stigmatization and exclusion whereby one becomes regarded as subhuman or inhuman.[20] All of the representations Doyle refers to linked Semenya's body and her comportment to her exceptional athletic ability, the assumption being that because she was so fast, so muscular, and so powerful, she couldn't *really* be a woman. The gendered ideology at work here is a familiar one for feminists: in a sweeping generalization that belies current empirical evidence, western journalists uncritically reiterated the belief that males are naturally more athletic, stronger, and faster than female athletes.[21] This ideology biologizes athletic capability and performance, renaturalizes sexual dimorphism, and normalizes masculinism as both the standard and the ideal of professional sport. Doyle importantly points out that "mainstream sports culture theatricalizes the exile and abjection of the feminine, the effeminate, the queer (isn't this what we mean by 'bullying'?). It stages gender segregation as not only natural but necessary to a sense of fairness. It does so in syncopation with a racialist logic that presents the black body especially as vitality, as raw force, as athleticism itself."[22] Rather than reading Semenya's win as a result of her time-honored commitment to athletic practice at the field and in the gym, of the daily grind of her hard work and dedication to training, western media focused on Semenya's so-called gender nonconforming appearance, which was then linked with her racialized status as a black South African woman. For the mainstream press in the global north/west, the primary "evidence" that Semenya's sex/gender was in question was presumed to be transparently visible on the surface of her body.

The media's sensationalistic treatment of Semenya predictably conflated sex and gender, which highlighted the widespread assumption that gender and sex both refer to the "natural," supposedly universal division of humans into what Anne Fausto-Sterling once called, tongue in cheek, "a perfectly dimorphic species."[23] On the other hand, some journalists' accounts inadvertently called this conflation into question, revealing enduring confusions over what the difference between sex and gender might be, where that difference might be located, and what its relationship to racial difference might be. But rather than sorting out the relationship between sex, gender, and race, journalists' largely sensational approach to these categories worked to muddy the already muddy waters. Indeed, media reports attempted and necessarily failed to manage and contain Semenya's narrative, which turned out to be anything but her *own* narrative, but rather a series of narratives written about her by both dominant regimes of power and critics of those regimes.

Clarey of the *New York Times* reports that Pier Weiss, the general secretary of the IAAF, was quick to stress that the sex test had been initiated because of "ambiguity, not because we believe she is cheating." In suggesting that it is ambiguity, not cheating, that is at stake for the IAAF, Weiss frames the two as completely separate matters. Yet there is an important sense in which Semenya's "ambiguity" became, in the discourses of the IAAF and the popular press, precisely a figure for cheating. The implication of the IAAF's position is that female athletes with ambiguous sex/gender presentations may have what some commentators have called an "unfair masculine advantage" and in that regard would be considered cheaters. I return to the utterly suspect concept of female athletes with masculine advantage shortly.

First I want to quote Clarey's article at length to provide some further context and also to examine at the ways that IAAF officials and fellow competitors responded to Semenya's win. According to Clarey,

> Weiss said that . . . if the investigation proves Semenya is not a woman, she would be stripped of the gold and the other medalists elevated. The investigation could take weeks, he said.
>
> "But today there is no proof and the benefit of the doubt must always be in favor of the athlete," he said. "Which is why we had no reason, nothing in our hands, to forbid the athlete to compete today."
>
> Not all of the finalists agreed. "These kind of people should not run with us," Elisa Cusma of Italy, who finished sixth, said in a postrace interview with Italian journalists. "For me, she's not a woman. She's a man."
>
> Mariya Savinova, a Russian who finished fifth, told Russian journalists that she did not believe Semenya would be able to pass a test. "Just look at her," Savinova said.[24]

In the first quotation Clarey provides, Weiss plays down the IAAF's investigation, asserting that the "benefit of the doubt must always be in favor of the athlete," even as his handling of the IAAF's investigation and its improper disclosure to the international press clearly violated Semenya's right to individual privacy, especially considering the fact that the results of her tests were, at the time he was interviewed by the *New York Times*, still pending.

Meanwhile, Cusma's remarks foreground the discriminatory attitudes that often emerge in response to persons who are seen to complicate dominant understandings of sex, gender, and their presumed relation. In her remarks, Cusma positions Semenya as a particular "kind" of person, the kind of person who should not compete in women's sport because, according to Cusma, "she's not a woman." Semenya's body becomes for Cusma a source and site of sex/gender trouble. At the precise moment when Cusma announces with unshakeable conviction a truth claim about Semenya ("She's not a woman. She's a man."), the sex/gender trouble that Semenya becomes a figure for comes to infect and permeate Cusma's own language, disclaiming the certainty she declares. That is, Cusma uses the feminine pronoun to refer to Semenya at the very same time that she denies Semenya's womanhood and femaleness. Both sentences, "she's not a woman" and "she's a man," refute the very claim that they assert. Or rather, what these sentences ultimately assert is that there is no referential or ontological guarantee that gendered pronouns necessarily refer to pre-given sexed morphologies. What is at stake here is not so much a body that resists the rhetoric of dimorphic determinism, although that may perhaps be at stake too, but, more to the point, the incoherence internal to the rhetoric of dimorphic determinism itself. Cusma's rhetoric thus sustains the very confusion over Semenya's sex/gender that Cusma claims to have seen through.

As my language indicates, the politics of vision played a fundamental role in the Semenya controversy, as also revealed by Savinova's comment, "Just look at her." As in the case of Cusma's remarks, the grammar of Savinova's sentence affirms the very proposition its rhetoric denies, naming Semenya with the feminine pronoun while simultaneously disclaiming Semenya's legitimacy as a woman. But, more than this, Savinova's comment reveals the fallibility of the widely held suspicion that a person's status as male or female, or masculine or feminine, can be plainly and objectively ascertained by one person's perceptual apprehension of another person's body. What is seeable depends as much if not more on who's looking, and through which cultural lenses, as it does on the object of the gaze. As Jasbir Puar argues, "the act of seeing is simultaneously an act of reading, a specific interpretation of the visual. But this reading passes itself off as a seeing, a natural activity, hiding the 'contestable construal' of what is seen."[25]

Following Puar's analysis of underacknowledged complexities of intersectionality as a visualizing technology, I argue that mainstream western discourse on Semenya was constructed through not only a gendered lens, but also a racial and a national lens. Savinova's comment "just look at her" may initially seem to refer to Semenya's sex/gender nonconforming appearance, but it can also be read as referring to Semenya's status as a black South African female. Likewise, it is important to notice the racial and national resonances of Cusma's comments, cited earlier: "For me, she's not a woman. She's a man." When Cusma makes this claim, she speaks not simply for herself, but as an Italian, a person of presumed Caucasian appearance, and as an official athletic representative of Italy and of Europe. Cusma thus marks Semenya's racialized sex/gender difference as uniquely susceptible to Eurocentric neocolonial visual scrutiny. As Pamela Scully argues, fellow athletes' and the media's treatment of Semenya speaks to "societal-wide tensions around varied and conflicting expectations of femininity, and how they are immersed also in racial representations. While white women runners have also faced challenges to their gender, black women have to navigate a more complicated terrain of racial and gendered domination which rendered and continues to render black female bodies particularly subject to the public gaze."[26]

Attending to the construction of racialized sex/gender in the politics of representation becomes crucial when we consider the materialization of what John Berger called "ways of seeing."[27] Ways of seeing are culturally naturalized and normalized modes of interpretation. The ways of seeing that focalize Semenya's gender as "ambiguous" call up and depend on not only the naturalization of sexual dimorphism, but also the reproduction of colonial ideologies that position racialized gendered subjects as simultaneously other to the normativity of white western femininity and as passive objects of the western biomedical gaze. In "Situated Knowledges," Donna Haraway contends that the field of vision does not merely reflect but rather is constituted by the technologies of seeing and structures of material-semiosis—the physiological, ocular, neurological, linguistic, cultural, and ideological cues and systems—that condition it.[28] From this perspective, Savinova's comment, "Just look at her," needs to be contextualized within a larger colonial history of vision and visibility in which black and brown female bodies and gender nonconforming bodies have long been treated as extraordinary objects of biomedical scrutiny and biopolitical regulation. Savinova figures the truth of sex/gender as if it were transparently written on Semenya's racialized body. Under western eyes, to invoke Chandra Mohanty's foundational formulation for postcolonial and transnational feminist analysis, it is as if the act of looking at Semenya performs a racialized sex/gender test of its own.[29]

In all these instances, we have to ask: by whose standard is Semenya being judged? According to which ways of seeing and knowing does Semenya not qualify as a woman? As in all matters of sex, gender, race, and nation, things turn out to be far more complicated than they appear at first sight.

From Ambiguity to Disorders of Sex Development

In the realm of professional athletics, as Anne Fausto-Sterling notes, the sex/gender distinction has long been a source of anxieties over the proper norms and forms of bodies.[30] According to Stanley I. Thangaraj, sport has been constructed as an arena wherein ideologies about human differences—codified in terms of intersecting formations of sex, gender, race, sexuality, nationality, class, and ability—are both reinforced and challenged.[31] Faced with ambiguity and uncertainty, the IAAF enlisted biomedicine to produce clarity. On January 20, 2010, Gina Kolata of the *New York Times* reported that, in response to the Semenya controversy, the International Olympic Committee (IOC) convened a panel of medical experts to devise a "treatment plan" for athletes with ambiguous sex characteristics.[32] As Kolata explains,

> Athletes who identify themselves as female but have medical disorders that give them masculine characteristics should have their disorders diagnosed and treated, the group concluded after two days of meetings in Miami Beach. The experts also said that rules should be put in place for determining an athlete's eligibility to compete on a case-by-case basis—but they did not indicate what those rules should be.

In an effort to maintain the neoliberal illusion of a fair playing field, the IOC singles out "athletes who identify themselves as female but have disorders that give them masculine characteristics." The IOC posits that certain so-called sex disorders give female athletes "masculine characteristics," but they do not take a similar position on athletes who identify as male and have disorders that give them "feminine characteristics," because feminine characteristics are not considered advantageous in most sporting disciplines. The IOC normalizes masculinity as both the standard and the ideal of athletic competition. Assuming that masculinity is composed of a set of natural physical qualities; that masculinity naturally accrues to certain bodies and not to others; that sex follows a natural course of dimorphic development; and that what defines females is their natural difference from males, the IOC naturalizes gender *as* sex. Moreover, in proposing that atypically sexed female athletes must be treated in order to compete, the IOC's decision further entrenches a transnational biopolitical regime of gender regulation that impacts athletes around the globe, wrapping up a social decision about what counts as sex, as Alice Dreger puts it, "as if it is simply a scientific decision."[33]

The IOC's invocation of the word "disorders" is therefore significant in this context, as it draws on what has come to be known as DSD nomenclature, a lexicon proposed in 2005 by medical experts as a replacement for the term intersex, a term which these experts argue has caused too much confusion—another word for gender trouble?—among parents and doctors of affected parties.[34] The proponents of the shift from intersex to DSD adopt medicalized language in an effort to generate improved treatment outcomes, but the DSD nomenclature has generated significant opposition from some intersex activist groups, such as Organisation Intersex International, whose members argue that the DSD nomenclature has discriminatory social and political implications. I explore the biopolitics and bioethics of DSD in greater detail in the conclusion to this book. Here the point I wish to make is that by using the Semenya controversy as a vehicle to sort out international policy on "athletes who identify themselves as female but have disorders that give them masculine characteristics," the IOC invited consideration of a) how the Semenya controversy is potentially linked to intersex, trans, and gender nonconforming issues; b) the ways that the rules of international athletics renaturalize specifically western conceptions of sexual dimorphism; and c) how these processes differentially play out in global northern and global southern contexts.

In 2009, without Semenya's consent, journalists with sources close to the IAAF investigation disclosed that the medical tests revealed that she had elevated testosterone levels and perhaps undescended testes.[35] Whatever the specific biochemistry and physiology of Semenya's body (which is *absolutely* none of anyone's business), Dreger argues that the IAAF's decision to subject Semenya to sex testing was disingenuous because the organization "has not sorted out the rules for sex typing and is relying on unstated, shifting standards."[36] What precisely counts as a disorder of sex development, and which authorities get to say? "The biology of sex," Dreger continues, "is a lot more complicated than the average fan believes." While many people believe that the biology of sex is a simple matter of XX or XY chromosomes, genitalia, and reproductive organs, Dreger importantly emphasizes that what we call "sex" has no single, unitary, overarching biological cause. As Dreger tells Clarey, "At the end of the day, [the officials] are going to have to make a social decision on what counts as male and female, and they will wrap it up as if it is simply a scientific decision . . . And the science actually tells us sex is messy."

Although Dreger is right to foreground the social overdetermination of the IAAF's policies on sex testing, her analysis does not address the racial and transnational dimensions of this "social decision." The social decision of what counts as male and female in the Semenya controversy was not only shaped by western biologizing discourses of sexual dimorphism but also by racialized logics and struggles over who counts—and can count—as a representative citizen within and across different national contexts.

In a 2009 article published in *The Nation,* sports writers Dave Zirin and Sherry Wolf link the Semenya controversy explicitly to the history and politics of intersex, arguing that what fans and officials still don't understand, "or will not confront, is that gender—that is, how we comport and conceive of ourselves—is a remarkably fluid social construction. Even our physical sex is far more ambiguous and fluid than is often imagined or taught."[37] In foregrounding the ambiguity of physical sex, Zirin and Wolf draw on intersex activist arguments to highlight the political nature of the sex/gender divide in professional sport. Moreover, they call attention to the human rights of intersex athletes as a key issue raised by the Semenya controversy. Why should intersex athletes—who are, after all, born with "natural" genetic, biochemical, and anatomical characteristics—be singled out for regulation, when athletes who might be cycling PEDs (performance enhancing drugs) are infrequently subjected to the same tests?

Although they provide a thoughtful critique of sex testing in international sport, Zirin and Wolf seem to presume that Semenya can be clearly and transparently considered an athlete with an intersex condition. However, as Zine Magubane stresses, "Semenya was never officially declared (nor did she declare herself to be) a person with an intersex condition."[38] "Semenya has always maintained her status as an authentic and authenticated female."[39] These claims suggest that reading the Semenya controversy solely through a western intersex activist lens may obscure additional complications at stake in the controversy. As Magubane suggests, to understand the Semenya controversy, we also need contextualize her situation in relation to South Africa's history as a former British colony. As I argue in chapter 4, intersex is a culturally and historically particular category, yet one that has been globalized by western biomedicine and transnational intersex activism in uneven ways. It is thus imperative to attend to the unique national histories and racialized sex/gender systems that have given intersex different meanings in the United States and in South Africa in the past and present.

Colonial Legacies, South African Black Nationalism, and the Reproduction of Sex/Gender Binarism

While Semenya was sensationalized by the mainstream western press, on the one hand, and defended as a victim of intersexism and biased sex testing by writers like Dreger and Zirin and Wolf, on the other, her reception by South African fans, athletic officials, and politicians was markedly different. As Magubane recounts, when Semenya returned home to South Africa after the controversy erupted in Berlin in 2009, she was greeted by majority black crowds holding signs that exclaimed, "Marry Me," "Welcome Home Our First Lady of Sport," and "100 Percent Female Woman."[40] South African athletic officials and politicians argued that the sex/gender tests Semenya was forced to undergo were an instance of "racism pure and simple."[41] "I'm angry, I'm fuming . . . You cannot say somebody's child is not a girl," Lenard Chuene, the president of Athletics South Africa at the time, told the *Los Angeles Times*.[42] Challenging the western media's depiction of Semenya as a gender outlaw, the president of South Africa, Jacob Zuma, said that she "showcased women's power and achievement"[43] and christened Semenya "our golden girl.[44]" These claims positioned Semenya as not only as a normatively sexed/gendered subject and proud symbol of South African athletic excellence and national pride, but also as an icon of postcolonial resistance to western domination. The South African politician Julius Malema interpreted the IAAF's treatment of Semenya in precisely these terms when he declared, "Don't impose your hermaphrodite concepts on us!"[45]

These examples reveal that South African athletics officials and politicians responded to the Semenya controversy by reproducing sex/gender binarism as part of an effort to mobilize Black Nationalism as a challenge to western imperialism. According to Neville Hoad, Malema was responding to the "shameful history of Sarah Baartman who was literally cut up and turned inside out for the world to see."[46] Analogizing Semenya's treatment with that of Baartman (pejoratively known as the Hottentot Venus), Hoad emphasizes the intersecting eugenic and sexological legacies of colonialism that have shaped discriminatory and exoticizing western attitudes toward colonized women's bodies. At the same time, drawing on queer theory, Hoad also critiques South African political leaders for "shrilly insisting on binarized gender as truly African."[47] Likewise, Tavia Nyong'o suggests that Semenya's defenders are not simply concerned with her well-being, but rather have an unmarked "patriarchal investment in naming and controlling this gender excess."[48]

Magubane argues that Hoade's and Nyong'o's critiques of Semenya's defenders implicitly rely on a western progress narrative that posits acceptance of sexual and gender diversity as an automatic sign of liberal modernity. To understand why Semenya herself, as well as many of her South African defenders, rejected the identity label intersex, Magubane contends that critical attention needs to be paid to the ways in which the intersex concept is overdetermined by race, imperial history, and national context.

Intersex, Race, and National Context

Not surprisingly, western scholars writing about Semenya tended to agree that her treatment reaffirmed the need for intersectional approaches to subjectivity and power. According to Nyong'o, "If ever a case called for an intersectional analysis that included queer and trans perspectives, as well as anti-racist and anti-imperialist ones, this should be one."[49] Although he uncritically uses the medicalized and objectifying word "case" to refer to the Semenya controversy, Nyong'o nonetheless makes evident that no single element of subjectification and no single form of power alone can explain the causes and effects of the IAAF's decision to subject Semenya to sex testing and the differing national responses to that decision. Brenna Munro concurred. "The Semenya affair," she writes, "underlines the importance of intersectional analysis informed by queer theory within African Studies. One cannot make sense of this spectacle without thinking about the afterlife of imperialism under globalization, the international politics of race, and how models of sex and gender normativity are produced and circulated in this context—and these forms of normativity are intimately linked with questions of sexuality."[50] These claims suggest that an intersectional analysis that takes intersex into account in relation to racialized gender, national context, and colonial legacies can enable us to forge a more refined understanding of the Semenya controversy. While I agree that the Semenya controversy problematizes monological and monocausal accounts of difference, I also want to suggest that an intersectional approach that excludes intersex, trans, and gender nonconforming folks is insufficient for understanding the politics of subjectivity, power, and representation that ultimately sought to contain Semenya's story.

Magubane argues that reading the Semenya controversy only as a case of bias against intersex athletes misses crucial factors that shaped the dominant iterations of the narrative in media, activism, and scholarship. In a comparative reading of twentieth-century US and South African medical texts on intersex, Magubane observes that "white and black intersexed bodies posed different kinds of threats in the United States and South Africa because of the different stance each nation took toward the potential for black participation in the common social and institutional life of the nation state. When we consider how intersex cases became visible in each locale, we must disaggregate by both nation and historical moment."[51] Magubane continues,

> The trauma of surgical correction arose to address a gender panic that was racially and nationally specific. Surgical correc-

> tion figured so prominently in the United States because of a political imperative to establish the normality of whiteness. An ambiguously gendered white body needed to be corrected to retain its whiteness, whereas an ambiguously gendered black body was seen as confirming the essential biological difference between whites and blacks.[52]

According to Magubane, the medicalization of intersex in the United States was motivated primarily by the ideological effort to maintain the whiteness of sexual dimorphism and binary gender. "By the time Money devised his optimal gender of rearing model and the concept of gender . . . the masculine and feminine states of being to which the concept referred were already and inherently coded as white and in opposition to blackness."[53] Money's concept of gender, she concludes, "had an exclusionary racial impulse written into it at its very inception."[54] Magubane's lucid analysis emphasizes that genealogies of intersex differ across national contexts. In the United States and South Africa, respectively, these genealogies were forged through different racialized logics of citizenship. Magubane thus helps us to understand why some of Semenya's defenders resisted "the sweeping imposition of a culturally specific category—intersex—on a culturally embedded individual."[55]

Although I agree with Magubane that national context, racialization, and imperial history overdetermine the visibility and subsequent management of bodies in doubt, her analysis overlooks the intersectional complexities of the uneven globalization of intersex medical management and intersex activism. This is perhaps because Magubane focuses on comparative national archetypes. According to Magubane,

> In the twentieth- and twenty-first century United States, the archetypal intersex person, who might have been subject to surgical correction, gender reassignment, and John Money's optimal gender of rearing model, had become a white child or adolescent (possibly one who goes on to become an adult activist). In South Africa, the medical and legal literature only started talking about intersex conditions in the 1950s. There, the archetypal patient has been a black adult who experienced neither involuntary gender reassignment nor surgery.[56]

While Magubane's comparison is illuminating, by limiting her analysis to an archetypical framework, she obscures the racial and ethnic diversity of patients subjected to intersex medical normalization in United States and also in other locations around the globe. Intersex—as a category of medicalized experience—is hardly an issue exclusive to white and middle-class US populations, as intersex activists of color such as Lynnell Stephani Long, Darlene Harris, Sean Safia Wall, and others interviewed in documentary films such as *Intersexion* and *One in 2000* make clear.[57] Moreover, as Cynthia Krauss's research on the medical normalization of intersex patients in West Africa suggests,[58] and as Mauro Cabral has argued, a significant number of intersex folks in the global south of diverse ethnicities have also been forced to undergo medical normalization.[59] As Cabral's critique of the coloniality of western normalizing technologies of gender regulation indicates, the medicalization of intersex was a project with globalizing ambitions, unevenly institutionalized and diversely stratified. In "Race and the Intersexed," Long observes that people of color in the global north/west have perhaps been especially reluctant to openly discuss the medicalization of intersexuality because they are already stigmatized in terms of both race and gender. This may also help to explain the raced, classed, and queer absences that Hilary Malatino identifies in western sexological archives, as I discuss in chapter 1.[60] It is very difficult if not altogether impossible to gather data on the racial and ethnic makeup of intersex patients globally. This is because medical providers are not required by law to record how many normalizing intersex surgeries and hormonal treatments they perform per year. According to research by Fausto-Sterling, intersex anatomical variation can be found across divisions of race, ethnicity, class, sexuality, ability, and nationality.[61] In this regard, Magubane's otherwise astute analysis obscures the fact that intersex normalization surgeries weren't exclusively performed to secure whiteness in the global north. They were also performed to reproduce the globalizing colonial ideologies of westocentric sexual dimorphism and binary gender within and across racially, ethnically, and nationally diverse populations.

Conclusion

To read Semenya's story critically requires recognizing that different constituencies—the western press; international athletic governing bodies; South African fans; athletic officials; politicians; and feminist, queer, intersectional, and postcolonial scholars—had different investments that shaped their various interpretations of her story. Although I do not believe that there is only one right way to read the Semenya controversy, I am convinced that scholars can produce more capacious critical readings by attending to how diverse scientific, social, political, and intellectual investments converged and diverged to shape the different interpretations of her story. While Magubane concludes that the Semenya affair challenges us to reconsider the overdetermination of intersex by race, imperial legacies, and national context, Sheri L. Dworkin, Amanda Lock Swarr, and Cheryl Cooky suggest that the Semenya controversy "pushes us to reconsider some of the strongly held premises of intersectionality, as it has been applied. What if categories such as race, class, gender, and nation do not operate equally to co-constitute one another but can actually work to obscure each other?"[62] With this vital question, Dworkin, Lock Swarr, and Cooky ask us to consider how the privileging of some categories of analysis may inadvertently dissimulate others, and how this dissimulation is globally and locally overdetermined. Might these erasures constitute an underinterrogated but fundamental aspect of the workings of biopower?

This question brings me back to Markowitz, who surely is correct that thinking beyond sexual dimorphism is not a utopian, impossible feat. However, dimorphism does not simply become superficial when we begin to grasp the genealogy of racialized gender, as my critical reading of Magubane suggests. To claim that the racialization of sex/gender renders binarism insignificant is to miss the ways in which binarism has been reproduced, reinscribed, and contested within and across the racial/sexual orders of diverse national and transnational contexts. It is also to obscure the place of the intersexed in genealogies of intersectionality, biopolitics, and geopolitics more broadly.

While Magubane is understandably critical of western intersex activist and scholarly interpretations of the Semenya controversy, she curiously does not attend to the perspectives and voices of South African intersex activists in her analysis. The late Sally Gross, the founder of Intersex South Africa, had this to say about the Semenya controversy and its complicated relationship to human rights and broader struggles for social justice:

> The support for Semenya from both people and government [in South Africa] offers real hope that things can be different. Some of this support is dismissive of the possibility that Semenya is intersexed. She may well not be intersexed; but the point which people and government surely need to take to heart is that even if she is intersexed, it should not be a "big deal." The intersexed are no less to be supported and protected against assaults on their human dignity than are people who are not intersexed. The people and government need to take up the plight of the intersexed no less vigorously than the cause of Caster Semenya. Government needs consciously to drive a process to ensure that the intersexed—infants, children and adults—are protected from

> assaults on their bodies, on their dignity as human beings and on other fundamental human rights.[63]

As Gross implicitly acknowledges, intersectionality and postcolonial critique can be enhanced by considering the place of intersex in contestations over what counts as intelligible human life within and beyond the South African context. Intersex has never been an issue extricable from processes of racial/sexual formation. As Gross's account as well as analyses by Krauss and Lock Swarr suggest, the claim that the category "intersex" is a pure imposition in the South African context obscures the longer and more complicated indigenous and colonial genealogies of South African sex and gender variation.[64] Although I critiqued the limitations of human rights approaches to intersex in the last chapter, here I find myself in agreement with Gross that a human rights perspective can still provide useful resources for re-situating intersex as *one among many* issues at stake in debates about differently situated bodies subject to doubt. While the Semenya controversy became a lightning rod for competing transnational claims about intersex, on the one hand, and intersectionality, on the other, Semenya's story cannot finally be reduced to either one. By highlighting the tensions between these claims and analyzing their convergences, divergences, exclusions, and erasures, this chapter has attempted to think critically about the myriad symbolic and material structures that shaped Semenya's story.

Semenya is still writing her story today, even as hegemonic institutions continue to attempt to appropriate her story for their own agendas. During the lead-up to the 2016 Olympics in Rio de Janiero, the controversy over Semenya was reignited once again. Although the IAAF cleared Semenya to return to competition in 2010, media reports about her road to the Olympics reiterated the same old tired colonial logics of gossip and speculation from the period in which, in her words, she was

> subjected to unwarranted and invasive scrutiny of the most intimate and private details of my being . . . [that] infringed on not only my rights as an athlete but also my fundamental and human rights including my rights to dignity and privacy.[65]

Katrina Karkazis connects the violence of the public scrutiny of Semenya to the media's unethical treatment of Indian sprinter Duttee Chand.[66] According to Karkazis,

> too many in the Global North have not appreciated the ramifications of being investigated and scrutinized for the livelihood

> of the individuals in question, their relationships, their freedom of movement, and their safety. The resulting stigma can lead to social ostracization, threatening career loss and decreased wages, the ability to marry, and one's deeply felt sense of self.[67]

As Semenya's words and Karkazis's interpretation of them make clear, marginalized subjects' rights to bodily integrity and self-determination are undermined by racialized practices of sex/gender regulation. As Semenya emphasizes, these practices contravene "fundamental and human rights," including the rights "to dignity and privacy."

What is most stunning, perhaps, is that Semenya's own interpretation of her experience has received far less attention than it deserves in media, political, activist, and scholarly discourses. In the following quotation, Semenya recognizes that she became a figure for other people's narratives, theories, and political desires. Moreover, she offers a subtle critique of the ways in which those narratives, theories, and political desires were imposed on her in ways that significantly constrained her existence. Although some readers might be tempted to render her words in terms of a neoliberal "born this way" discourse, what such a reading and most accounts of Semenya's story generally miss is precisely the resistance to imposed narratives that she expresses at multiple levels of her being/becoming—in and through her athletic performance and her speech acts. It is difficult to say whether Semenya's resistance springs from oppositional and differential modes of consciousness, whether it bears the material-semiotic traces of subjugated knowledge.[68] In any case, one thing is clear: like countless historically dispossessed subjects, Semenya's story was never entirely her own.[69] It is therefore fitting that, in the context of this chapter, Semenya herself should have the last word:

> I am not a fake. I am natural. I am just being Caster. I don't want to be someone I don't want to be. I don't want to be someone people want me to be. I just want to be me. I was born like this. I don't want any changes.[70]

Conclusion

Thinking Intersex Otherwise

Disorders of Sex Development, Social Justice, and the Ethics of Uncertainty

> What if we were to reach, what if we were to approach here (for one does not arrive at this as one would a determined location) the area of a relationship to the other where the code of sexual marks would no longer be discriminating?
>
> —Jacques Derrida[1]

This book concludes with an analysis of one of the most recent shifts in intersex politics: the western biomedical establishment's effort to replace the term *intersex* with the taxonomy *disorders of sex development* (DSD). I situate the DSD debates in the context of contemporary reterritorializations of empire, processes of neoliberal restructuring, and the retooling of biopolitical technologies of bodily regulation. By doing so, I hope to show one last time that queer feminist science studies, transnational feminisms, and intersectional perspectives provide vital resources for articulating ethical alternatives to the medicalization of people of atypical sex and, ultimately, for thinking intersex/DSD otherwise.

In July 2008, the Intersex Society of North America closed up shop. The factors underlying ISNA's dissolution were multidimensional and complex. While over its fifteen-year history the organization pursued many successful initiatives to raise awareness about the medical treatment of infants and persons with intersex conditions, by the early 2000s a number of ISNA board members, including its founder and longtime executive director, Cheryl Chase, began to feel that the organization's previous politicized activist efforts had not been effective enough. A new approach was needed.

That new approach was unveiled to the world at a 2005 conference held in Chicago hosted by the Lawson Wilkins Pediatric Endocrine Society and the European Society for Pediatric Endocrinology. At the conference, a range of medical specialists in pediatrics, endocrinology, urology, and other medical disciplines gathered alongside representatives from ISNA to publicly announce that the term *intersex* had outlived its usefulness. In place of *intersex*, they proposed a new term, *disorders of sex development*, referred to by the acronym DSD. The image of intersex activists standing side by side with medical professionals to jointly announce DSD as a better replacement for intersex may seem striking, even perplexing, to those familiar with the intersex movement's genesis. After all, intersex activism emerged in North America and elsewhere around the globe in the 1990s as an oppositional critique of the medial pathologization and normalization of corporeal variation in the name of social harmony. It is therefore worth querying the assumptions underlying and the repercussions of this shift, which is simultaneously a shift in language, medical practice, and activist strategy.

A series of events immediately followed the Chicago conference that sought to legitimize and institutionalize DSD as a hegemonic category. First, in 2005 I. A. Hughes et al. published a "Consensus Statement on Management of Intersex Disorders" in the *Archives of Disease in Childhood*, the premier pediatric medical journal, which argued that the DSD nomenclature avoids the confusion and stigma associated with the term *intersex* and will thereby facilitate improvements in clinical care.[2] Second, in 2006 historian of medicine and then-current member of ISNA Alice Dreger, under the auspices of an organization called the Consortium on the Management of Disorders of Sex Development, published the twin pamphlets "Clinical Guidelines for the Management of Disorders of Sex Development in Childhood" and "Handbook for Parents."[3] Working in partnership with a new nonprofit organization called Accord Alliance, which opened its doors in 2008, the Consortium adopts medicalized language and strategies to approach DSD treatment in terms of what they call "patient-centered care." As the Accord Alliance mission statement explains, their goal is to "promote comprehensive and integrated approaches to care that enhance the health and well-being of people and families affected by disorders of sex development (DSD, which includes some conditions referred to as 'intersex')."[4]

Taking stock of these developments, historian Elizabeth Reis notes that medical providers, parents, allies, and some people with intersex conditions in the global north/west and in some parts of the global south/east have rapidly embraced the DSD nomenclature.[5] Proponents argue that the terminological shift is enabling physicians to focus on what Ellen K. Feder and Katrina Karkazis call "the genuine medical issues associated with intersex conditions," leading to more successful treatment outcomes, although no empirical data have been gathered to substantiate these claims.[6] Feder defends this shift in the following way: "Rather than fight for the demedicalization of intersex conditions that indeed have consequences for individuals' health, acceptance of this change can transform the conceptualization of intersex conditions from their past treatment as 'disorders like no other' to 'disorders like many others.'"[7] Acknowledging that all forms of sex development have multiply mediated and complex consequences for individuals' health, I nonetheless question Feder's logic. Simply put, I am not convinced that sex is a category like any other. As a category of difference, sex is biologized as not only dimorphic but also, and crucially, embryonic and molecular in a way that other categories of difference are not. Sex is the first question we ask of expectant parents and the caretakers of newborns alike. Dimorphic sex is so deeply naturalized that it has become a *primal* and *primary* grid or schema through which an infant becomes intelligibly human, even as it is not the only such grid. This is why, after all, the birth of an infant with a sexual anatomy deemed to be nonstandard remains troubling for many parents and doctors. Feder's argument for normalizing disorders of sex development as "disorders like many others" depends on the assumption that sex and disability are analagous categories of difference, rather than intersectional ones. Moreover, Feder fails to question the biopolitical gatekeeping structure that is reproduced when biomedicine conspicuously assumes the role of chief arbiter for which sexed bodies count as disordered and which do not.

Drawing on certain neoliberal currents within disability studies and beyond, Feder frames the term "disorder" as inherently destigmatizing. This is a perplexing move for a self-described Foucauldian.[8] As Feder's usage of the related terms "health" and "conditions" also suggest, she naturalizes medicalized language, assuming that its referents are static, objective, empirical givens rather than contested biocultural materialities forged in and through the politics of difference. Feder advocates that medicine should expand the conditions of "human flourishing"[9]—another term she leaves uninterrogated[10]—for all segments of the population. She therefore interprets the adoption of DSD as "progressive" because it will supposedly allow medical practitioners to shift away from the concern with maintaining normative gendered appearance toward a focus on the "genuine health challenges faced by many individuals with intersex conditions."[11] In theory, this sounds like a forward-thinking perspective. But can we so easily separate "genuine health challenges" from regulatory schemas of gendered embodiment? What structures determine the difference between genuine health challenges and, apparently, disingenuous ones? Furthermore, which people's voices and experiences are erased and foreclosed by reading DSD as a more legitimate and even "healthier" name for bodies that were previously labeled intersex? In short, which forms of power and authority are reconsolidated through the shift to DSD?

These questions are at the heart of a growing transnational critical opposition to the DSD nomenclature. As Georgiann Davis brilliantly argues, "medical professionals took so quickly to the new DSD nomenclature because it allowed them to escape their tainted history of intersex medical care. With medical authority and jurisdiction over the intersex body in jeopardy, the new language allowed medical professionals to reassert their power over intersex."[12] Lending support to Davis's analysis, Morgan Holmes and Jennifer Germon use feminist, queer, and crip theory to call attention to how the DSD nomenclature renaturalizes the normal/pathological distinction and thereby reproduces sexual dimorphism, binary gender, heteronormativity, and compulsory able-bodiedness as intersectional ideological fascia of human intelligibility.[13] As Reis, who also draws on crip theory in her analysis, explains, "the word *disorder* connotes a need for repair," and this new nomenclature thus "contradicts one of intersex activism's central tenants: that unusual sex anatomy does *not* inevitably require surgical or hormonal correction."[14] The term *disorder* presumes that sex has a natural, proper, and "ordered" form of dimorphic development in the first place. Relabeling intersex as DSD reinforces the idea that variations of anatomical sex are less livable and less acceptable; and this relabeling naturalizes non-intersex anatomies as both normal and socially desirable.[15] Shifting from *intersex* to *disorders of sex development* thereby denies the fluidity and variability of sex, gender, and disability across historical and cultural contexts.[16]

In a related vein, Organisation Intersex International (OII), the largest intersex activist alliance in the world, with affiliates in twenty countries on six continents, has articulated a social justice critique of medicalization that shares similarities with Chase's earlier, more radical queer feminist views that I address in chapter 3.[17] Rejecting the premise that intersex bodies are inherently disordered, OII argues that the rhetoric of *disorders* not only repathologizes intersexuality, but also extends eugenic legacies, positioning human anatomical diversity as perpetually in need of specialized disciplinary medical management. In addition, OII has implicated the shift to DSD in contemporary reterritorializations of global northern/western empires. OII "strongly opposes current scientific terminology such as 'disorders of sexual development,' arguing that it is reductionist and imperialist."[18] As I noted in chapter 1, OII's intersectional critique of the DSD nomenclature holds that DSD extends rather than interrupts western imperialism's xenophobic, racist, sexist, homophobic, and transphobic legacies and its consequent maldistributions of life chances for disenfranchised populations. For OII, the medicalization of intersex is fundamentally linked with multiple, overlapping systems of power-knowledge. Their critique implicates western DSD *and* intersex rights advocacy within a larger web of historical, political economic, and ideological relationships—none of which is transparent or unmediated—and all of which must be analyzed in relation to larger histories of biopolitics, capital, and empire. This is why the politics of location of such activism, like all forms of world making, warrants critical examination. OII's question, in essence, is: what kind of a world are advocates of the DSD nomenclature *for*?

OII's question can be extended to patient-centered medicine.[19] In contemporary intersex/DSD rights advocacy, the patient-centered model is consistently hailed as a lynchpin for moving toward progressive medical reform in ways that will protect patients' human rights. Patient-centered care prioritizes "the person's right to make informed choices about their own bodies, and delaying treatment until the patient can make informed consent."[20] However, although patient-centered care mitigates the psychosocial rush to infant genital surgery, it does not rule out surgery as an option in the treatment of said disorders. Using the rhetoric of "rights," "choice," and "consent," proponents figure the patient-centered approach as an extension of US neoliberal democratic frames of subjectivity. According to Emily Grabham, the patient-centered model "uses the rhetoric of the patient as consumer: a service-user who expects a certain standard of care, and who should expect a range of treatment options and the opportunity to exercise their personal choice in making treatment decisions."[21] Grabham contends that this rhetoric conflates consumerism with citizenship in ways that have the potential to "re-embed material inequalities."[22] Grabham writes,

> The rhetoric on which [the patient-centered model] is based is, arguably, the neoliberal rhetoric of an autonomous intersex individual who, given the current socio-medical context, simply does not exist . . . Gaining consumer citizenship on the back of postponing "consent" to intersex surgeries or other forms of non-medically necessary intersex treatments to later in the child's life, as advocated by the patient-centered model, does not fundamentally challenge the disciplinary function of medical constructions of sex. It does not contest the idea that the medical sphere is in fact the correct sphere for intersex issues to be negotiated. At some future point, a more radical non-interventionist strategy might work to contest the privileging of

> medical discourses in responding to intersex issues in the first place. Thinking through these issues illuminates the significance of a critical approach to rights in the intersex context.[23]

For Grabham, the central question is whether and how advocates of intersex/DSD rights might evade "the disciplinary effects of their strategic claims for consumer citizenship."[24] Crucially, the patient-centered model treats medical care as apolitical and thereby obfuscates the role that medicine has played and continues to play in upholding colonial, capitalist, heteropatriarchal, white supremacist, ableist, and other types of hierarchies.[25] With its emphasis on neoliberal modes of self-fashioning and personal responsibilization, the patient-centered model reifies the humanist fiction of the sovereign subject. It thus fails to recognize that patients' autonomy is always already regulated by "the disciplinary function of medical constructions of sex," as well as raced, classed, gendered, able-bodied, and other forms of health care disparities.[26]

Although patient-centered care appears to be a clear improvement over the "concealment model" of intersex management,[27] centering the patient without critically attending to the ideological, political, and economic forces that shape the unequal institutional distribution of access to and quality of care risks promoting an excessive individualism. It targets patients and their bodies, rather than the social worlds they inhabit, as the site in need of transformation. Like the DSD rubric, then, patient-centered care does not disrupt the pathologization of bodies that exceed and trouble normalized westocentric classificatory schemas. It does not contest the medicalization of intersexuality as such. Instead, patient-centered care recasts medicalization in neoliberal terms as a seemingly sensible "choice" for individuals born with bodies that do not conform to the mythical norm.[28] In this regard, patient-centered care and the DSD rubric warrant critique especially because they dissimulate the heteronormative, eugenic, and (neo)colonial legacies of biopolitics. That is, they assign unequal gradations of social value to different embodied states and presume to know in advance what kind of body one must have to qualify as intelligibly human.

The rapid global dissemination of the DSD nomenclature and the patient-centered model returns our attention the problem of "intersex imperialism" and also exposes the need to think critically about the Enlightenment "will-to-knowledge" that governs the biomedical sciences in an era of neoliberal globalization.[29] Crucially, throughout the DSD debates, relatively little attention has been paid to the ways in which the DSD nomenclature attempts to shift concern away from the psychosomatic stigma and trauma produced through the medicalization of intersex. It is my contention that

that the turn to DSD can be understood as a legacy and inheritance of, rather than a divergence from, Money's OGR paradigm. Like Money's approach, the DSD rubric constitutes a medico-scientific attempt to pin meanings and bodies down. And like Money himself, the DSD advocates invest in the scientificity of diagnostic categories, even though so much intersex scholarship and activism—not to mention queer, feminist, and postcolonial work in science studies—challenges the pretense that medical science is inherently objective, uninflected by politics, and invariably benevolent.[30] Ultimately, the institutionalization of DSD reiterates Money's belief in the possibility and veracity of a universal scientific language for talking about sex and his grounding assumption "that only with a common language can one share a world."[31] It is this assumption that must be contested in order to generate alternative intersex imaginaries.

The DSD advocates propose a change in nomenclature—a change in language—as a means toward improving patient care. The linguistic change they recommend is not about replacing one term with another that basically means the same thing; rather, it is about instituting a new term—DSD—with a different and highly specific medical meaning to control and cover over the historicity, semantic volatility, instability, and ambiguity of an older term—intersex. Instead of critically interrogating the sources and effects of the material-semiotic instability or flux of *all* bodily categorizations, the DSD proponents seem to presume that replacing intersex with a more "scientific" and less *troubling* term will ameliorate the uncertainties about sex and gender, and sociality and subjectivity, that intersexuality exposes. For the DSD advocates, then, the term intersex is no longer "useful" because it shakes up and confuses dominant ways of categorizing and understanding bodies. But this is precisely why I suspect that the term intersex will live on, alongside and perhaps even beyond contestations over DSD, because intersex—as a technology of subject formation, object of knowledge, analytic category, and mode of dis/identification[32]—troubles the notion that only with a common language or onto-epistemology can one belong in a world of differences.

From a perspective attuned to the ways in which the medicalization of intersex/DSD is part of a larger process that involves biopower's targeting of the field of embodied differences broadly construed, it becomes possible to critically query the institutionally vested and regulatory aspects of the DSD nomenclature. As my discussion of Caster Semenya in chapter 5 suggests, contra Feder's argument cited above, the language of pathology can substantiate and justify medical and social systems of sex/gender normalization, systems that force particular subjects to *prove* their femaleness or maleness to scientific and legal authorities, even to the point of undergoing treatment, in order to qualify as intelligibly human. Indeed, both the Semenya story and the turn to DSD expose the persistence of a medico-scientific desire to create docile bodies.[33] In the face of this powerful will-to-knowledge, how might we begin to cultivate and proliferate alternatives?

This book has argued that queer feminist science studies, transnational feminisms, and intersectionality provide indispensible resources for rethinking the centrality of intersex in the workings of biopolitics and geopolitics. I have also suggested that this theoretical orientation can help to foment an ethics that embraces the enduring uncertainties about embodiment that intersex bodies expose. Before concluding, I would like to expand on how this ethics of uncertainty contributes to the effort to think intersex otherwise.

Ethics of Uncertainty

Reis has suggested that the term *divergences of sex development* might be a suitably "non-stigmatizing, non-correction-demanding" replacement for *disorders of sex development*.[34] Although I appreciate the spirit of Reis's proposal, I have concerns about the acronymic interchangeability of her substitute (DSD) with the original DSD nomenclature, which remains a pathologizing rubric. To be sure, *divergence* means something quite different than *disorder*, but this difference risks being erased when alphabetically interchangeable acronyms can so easily substitute for shortcuts for critical thinking. In a similar yet different vein, Anne Tamar-Mattis and Milton Diamond offer the term *variations of sex development* (VSD) as a less culturally biased and more physiologically accurate description of the sheer heterogeneity of human anatomical diversity.[35] Although I remain skeptical of the ideological work performed by the reduction and reification of any keyword or set of keywords to an acronym within neoliberal contexts, what is useful about both alternatives is that they indicate the need to rethink how communities produce knowledge about sex/gender in everyday life on an ongoing basis. In other words, Reis and Tamar Mattis and Diamond suggest the need for a persistent critique.

Whatever new word affected parties, medical providers, parents, activists, and scholars decide to employ as a replacement for intersex—disorders, divergences, or variations of sex development, or some other term—we cannot afford to presume that any of these terms is neutral or transparent. All of these terms have specific, not unrelated genealogies. All have material-semiotic implications that require analysis. If and when we choose to use one or another of these words, it seems incumbent on us to acknowledge that this usage depends on the assumption of our prior acquiescence to the view that the word intersex was never able to withstand what was asked of it in the first place: to be a placeholder, container, or proxy for the transference of western anxieties about the uncertainties of embodiment.

The genealogies of intersex I have traced in this book are in many ways stories about the anxieties provoked by uncertainties. "Perhaps there are," Fausto-Sterling writes at the end of her book *Sex/Gender: Biology in a Social World*, "things about sex and gender that we can never know."[36] This is not a statement about what we know, or a statement about what we know that we don't know, about sex and gender. It is not a declaration of knowledge or its lack. Rather, it is a reminder about the import of acknowledging the uncertainty instituted by the "perhaps." *Perhaps* there are things about sex and gender that necessarily exceed the limits of our knowing practices, and we cannot say for sure one way or the other. Fausto-Sterling's formulation foregrounds uncertainty—the state of not being certain, of being in doubt, and of existing without epistemic or ontological resolution or closure—as integral to thinking through the *possible* limits of the study of sexed and gendered life.

Taking up Fausto-Sterling's insight, it is my contention that uncertainty can also be reconceptualized as an ethical orientation.[37] As Jacques Derrida suggests, uncertainty is paradoxically a condition of responsibility.[38] Thomas Keenan puts it like this:

> Responsibility is not a moment of security or of cognitive certainty. Quite the contrary: the only responsibility worthy of the name comes with the removal of grounds, the withdrawal of the rules or the knowledge on which we might rely to make our decision for us. No grounds means no alibis, no elsewhere to which we might refer in the instance of our decision. If responsibility has always been thought in the western ethical, political, and literary traditions as a matter of articulating what is known with what is done, we propose resituating it as an asymmetry or an interruption between the orders of cognition and action. It is when we do not know exactly what we should do, when the effects and conditions of our actions can no longer be calculated, and when we have nowhere else to turn, not even back onto our "self," that we encounter something like responsibility.[39]

At an ethical register, uncertainty reorients us toward the question of how to make decisions—how to act responsibly—in the face of the undecidable and the uncertain, which Derrida calls the space of freedom. In a world without absolute measure, decisions must nevertheless be made. Decisions happen in the "night of non-knowledge and non-rule."[40] They are made in the dark, without a rulebook, as it were.[41] Uncertainty, then, is not an impediment to responsible action but rather its precondition.

The ethics of uncertainty has roots in diverse traditions of critical thought and practice. Its antecedents lie not only in Derrida's and Keenan's work, but also in certain transnational feminist, queer, antiracist, intersectional, decolonial, trans, and crip modes of praxis that seek to keep the question of difference alive.[42] This ethical tradition also has a multifaceted genealogy in intersex studies. It can be traced in the early work of Chase; in scholarship by Kessler, Fausto-Sterling, Judith Butler, Susan Stryker, Iain Morland, Morgan Holmes, María Lugones, Julia Sandra Bernal Crespo, Mauro Cabral, Zine Magubane, and Hilary Malatino; in the activisms of Organisation Intersex International, Advocates for Informed Choice, and the youth group Inter/Act; and in the poetic, creative, and artistic productions of Thea Hillman, Del LaGrace Volcano, and Ins A Kromminga, among many others.[43] Starting from different points of departure and using a wide range of means without predetermined ends, these activists, scholars, artists, and cultural workers variously articulate conceptions of difference that seek to make a difference in the world. Rather than instituting a need to know the "truth" of difference or alterity in advance, these thinkers critique the injustices of the erasure of difference in the worlding of the world and simultaneously constructively work to imagine how the world might be worlded otherwise.

In the immemorial words of Audre Lorde, "Certainly there very real differences between us of race, age, and sex. But it is not those differences between us that are separating us. It is rather our refusal to recognize those differences, and to examine the distortions which result from our misnaming them and their effects upon human behavior and expectation."[44] In the specific context of Lorde's black, lesbian, and feminist philosophical and poetic archive, this quotation is irreducible to any soft multicultural endorsement of the value of difference for neoliberalism. Critically conscious of multiple, intersecting modes of domination that shape diverse people's lives, Lorde identifies a multilayered series of impediments to counterhegemonic world-making: first, "our refusal to recognize"—that is, to *re-cognize*, to critically rethink—the differences between us, what counts as difference, or the question of difference itself; second, our inability to "examine the distortions which result from our misnaming them and their effects upon human behavior and expectation," or the ideological mystifications that ensure the codification of difference as deviance, pathology, and inferiority; third, our inability to accept (to be hospitable toward) differences in their irreducible singularity and plurality; and fourth and finally, our reluctance to celebrate (to revalue) those differences, and thus to be open to the transformations such a revaluation might generate.

Drawing on Lorde's intervention, the ethics of uncertainty that I seek to animate reframes difference not as an obstacle to relationality but rather as a condition of ethical co-belonging in a world with others. By suggesting that alternatives to the medicalization of intersex/DSD can emerge from an ethics of uncertainty, I mean to emphasize that affected parties, doctors, parents, politicians, policy makers, activists, educators, and workers have the capacity to make different decisions, decisions that acknowledge the uncertain and the undecidable and thereby produce different epistemologies of embodiment and different forms of ethical relation. We could, for instance, recognize that many people are born with characteristics that do not neatly fit into hegemonic corporeal schemas, and we could welcome, affirm, and revalue their differences. This is why I argue, following Morland's perspicuous contribution, that an ethics of uncertainty offers a more conscientious platform for how to treat intersex and non-intersex individuals alike.[45]

Unlike the prevailing biomedical approach to intersex/DSD, an ethics of uncertainty does not concede that one particular configuration of the human species is inherently more valuable than others or that fitting in is always preferable to not knowing precisely where one fits in the first place. Better alternatives to the medicalization of intersex/DSD can be found in the efforts of diverse people to revalue a radical heterogeneity of embodied lives without claiming the need to know in advance the precise forms those lives should or could take. Such alternatives are immanent in the uncertainties acknowledged by Fausto-Sterling's *perhaps*, even as they remain yet to come.[46]

Thinking Intersex Otherwise

The existence of people for whom dominant sex and gender classification systems fall short brings into sharp relief the arbitrary nature and fundamentally biopolitical character of binary schemas of sex and gender. Making the world more habitable and more hospitable for people whose anatomies call intersexist, heteronormative, masculinist, transphobic, white supremacist, and able-ist technologies of gender regulation into question will require learning to see the bodies of others and ourselves—whether we identify as intersex or not—in new ways. In other words, it will require the ongoing critical revision and revaluation of the politics of embodiment on both local and global scales.

Thinking intersex otherwise is thus the flip side of the coin of an ethics of uncertainty. Thinking intersex otherwise entails recognizing, in Fausto-Sterling's words, "that bodies are not bounded."[47] Which is to say, bodies are not set in stone, dictated by fixed, timeless, intransigent, and unchanging natural and/or cultural laws. Rather, bodies change and adapt in response to their naturecultural environments (which are themselves always variably changing and adapting). This is why bodies don't come in only one form, nor do they end at the skin.[48] Embodiment is extraordinary. It is where bios meets ethos. To learn to be open to the boundless, the incalculable, and the uncertain—isn't this perhaps the most difficult and fundamental lesson of the intersexed?

Notes

Introduction

1. Judith Butler, *Undoing Gender* (New York: Routledge, 2004), 65.

2. Donna Haraway's concept of the material-semiotic refers to how processes of materialization and signification are not opposed or mutually exclusive, but rather are interactive and co-constitutive in non-totalizing and nontransparent ways. See Haraway, "Situated Knowledges: The Science Question in Feminism and the Privilege of Partial Perspective," in *Simians, Cyborgs, and Women: The Reinvention of Nature* (New York: Routledge, 1991), 183–202; and Haraway, *When Species Meet* (Minneapolis: University of Minnesota Press, 2007).

3. Some scholars place intersex within the trans and gender nonconforming umbrella, while others do not. I elaborate on these distinctions below. The relationship between critical intersex studies and trans studies is contested. See, for instance, Talia Mae Bettcher, "Intersexuality, Transgender, and Transsexuality," in *Oxford Handbook of Feminist Theory*, ed. Mary Hawkesworth and Lisa Jane Disch (New York: Oxford University Press, 2016), 407–27. Importantly, as Vernon Rosario notes, the line between intersex and non-intersex anatomies has begun to blur under the weight of growing scientific evidence. See Rosario, "Quantum Sex: Intersex and the Molecular Deconstruction of Sex," *GLQ* 15, vol. 2 (2009): 267–64; and David A. Rubin, "Biochemistry and Physiology," in *The Wiley Blackwell Encyclopedia of Gender and Sexuality Studies*, ed. Nancy Naples (New York: Wiley Blackwell, 2016), 1–7. In this book, I use the intersex/non-intersex distinction loosely, recognizing that even as biomedical institutions and social conventions continue to hold fast to this distinction, the distinction itself is becoming increasingly unstable and uncertain because of interconnected shifts in research, culture, and politics.

4. Vernon Rosario, "Quantum Sex."

5. The frequency of intersex is difficult to estimate for several additional reasons. First, there is the lack of reliable data. As Noah Ben-Asher points out, the one to two in two thousand figure does not include the many individuals born with *subtler* corporeal variations, such as Klinefelter syndrome, Turner syndrome, or vaginal agenisis, some of which do not appear or are not diagnosed until later in life. Second, because the category *intersex* encompasses a wide variety of

disparate conditions and anatomical differences that are neither causally nor necessarily correlatively linked—for instance, congenital adrenal hyperplasia (CAH), androgen insensitivity syndrome (AIS), hypospadias, 5-alpha reductase deficiency, and a variety of others—intersexuality is by definition an imprecise label, more a catchall than a precise diagnosis. Third and most importantly, the frequency of intersex depends on how intersex is defined. And how intersex is defined, Ben-Asher perspicuously observes, "depends on what counts as 'male' and what counts as 'female' in a given society at a given time" (80). See Ben-Asher, "The Necessity of Sex Change: A Struggle for Intersex and Transsex Liberties," *Harvard Journal of Law and Gender* 29, no. 1 (2006): 51–98.

6. Melanie Blackless, Anthony Charuvastra, Amanda Derryck, Anne Fausto-Sterling, Karl Lauzanne, and Ellen Lee, "How Sexually Dimorphic Are We? Review and Synthesis," *American Journal of Human Biology* 12, no. 2 (March/April 2000): 151–66.

7. I use the terms "global north" and "global south" in this book to refer to current geopolitical and planetary cartographies of difference and power, even as I recognize that these terms—like the terms "developed world/developing world" and "east/west," all of which are contested—are inadequate to mapping the complexity of global-local power dynamics in any particular situation. For an elaboration of the global north/global south distinction in the context of contemporary transnational intersex activisms, see Dan Christian Ghattas, *Human Rights Between the Sexes: A Preliminary Study on the Life Situations of Inter* Individuals* (Berlin: Heinrich-Boll-Stiftung, 2013).

8. See N. S. Crouch, C. L. Minto, L.-M. Laio, C. R. J. Woodhouse, and S. M. Creighton, "Genital Sensation after Feminizing Genitoplasty for Congenital Adrenal Hyperplasia: A Pilot Study," *BJU International* 93 (2004): 135–38; M. Diamond and H. K. Sigmundson, "Sex Reassignment at Birth: Long-term Review and Implications," *Archives of Pediatric and Adolescent Medicine* 151 (October 1997): 298–304; John Coltapinto, *As Nature Made Him: The Boy Who Was Raised a Girl* (New York: Harper Collins, 2000); Amy Bloom, *Normal: Transsexual CEOs, Crossdressing Cops, and Hermaphrodites with Attitude* (New York: Vintage, 2003); and Eric Parens, ed., *Surgically Shaping Children: Technology, Ethics, and the Pursuit of Normality* (Baltimore: Johns Hopkins University Press, 2006).

9. Sarah M. Creighton, Catherine L. Minto, and Stuart J. Steele, "Objective Cosmetic and Anatomical Outcomes at Adolescence of Feminising Surgery for Ambiguous Genitalia Done in Childhood," *Lancet* 358 (2001): 124–25.

10. Sharon Preves, *Intersex and Identity: The Contested Self* (New Brunswick: Rutgers University Press, 2003); Katrina Karkazis, *Fixing Sex: Intersex, Medical Authority, and Lived Experience* (Durham: Duke University Press, 2008); Georgiann Davis, *Contesting Intersex: The Dubious Diagnosis* (New York: New York University Press, 2015); Alice Dreger, ed., *Intersex in the Age of Ethics* (Hagerstown: University Publishing Group, 1999); and Morgan Holmes, *Intersex: The Perilous Difference* (Selinsgrove: Susquehanna University Press, 2008).

11. See Anne Fausto-Sterling, *Sexing the Body: Gender Politics and the Construction of Sexuality* (New York: Basic Books, 2000); and Davis, *Contesting Intersex*.

12. Anne Fausto-Sterling, *Sex/Gender: Biology in a Social World* (New York: Routledge, 2012).

13. See Barbara Ehrenreich and Deirdre English, *For Her Own Good: Two Centuries of the Experts' Advice to Women* (New York: Anchor Books, 2005); Dorothy Roberts, *Fatal Invention: How Science, Politics, and Big Business Re-create Race in the Twenty-First Century* (New York: The New Press, 2011); Kathy Davis, *The Making of Our Bodies, Ourselves: How Feminism Travels Across Borders* (Durham: Duke University Press, 2007); and Morgan Holmes, "Mind the Gaps: Intersex and (Re-Productive) Spaces in Disability Studies and Bioethics," *Bioethical Inquiry* 5, no. 2 (2008): 169–81.

14. See, for instance, the essays collected in Morgan Holms, ed., *Critical Intersex* (London: Ashgate, 2009).

15. Audre Lorde, *Sister Outsider: Essays and Speeches* (New York: The Crossing Press, 1984), 116.

16. Holmes, *Intersex: A Perilous Difference*, 54.

17. Anne Fausto-Sterling, "The Five Sexes, Revisited," *Sciences* 40, no. 4 (2000): 19–20.

18. Suzanne J. Kessler was one of the first feminist scholars to make this argument. See "The Medical Construction of Gender: Case Management of Intersexed Infants," *Signs* 16, no. 1 (1990): 33–38.

19. Eve Kosofsky Sedgwick, *Epistemology of the Closet* (Berkeley: University of California Press, 1991), 22.

20. Ibid., 22.

21. In this regard, *Intersex Matters* investigates how Sedgwick's distinction between minoritizing versus universalizing approaches to homosexuality might be applied to intersexuality. A minoritizing view suggests that intersexuality is an active issue of importance only for a relatively small portion of the population. By contrast, a universalizing or majoritarian view maintains that the politics of difference are, to borrow Sedgwick's formulation, an "unpredictably powerful solvent of stable identities" (85) and that intersexuality "is an issue of continuing, determinative importance in the lives of people across" the spectrums of sex, gender, sexuality, race, ability, and nation (1). Adopting a majoritarian perspective, I contend that intersex matters hold relevance not merely for a small subset of the population, but rather for all subjects whose experiences are shaped by biomedical technologies of health and normalization. See Sedgwick, *Epistemology of the Closet*, 1–66.

22. Cheryl Chase, "Affronting Reason," in *Looking Queer: Body Image and Gay Identity in Lesbian, Bisexual, Gay, and Transgender Communities*, ed. Dawn Atkins (New York: Harrington Park Press, 1998), 204; reprinted in Joan Nestle, Clare Howell, and Riki Wilchins, eds., *GenderQueer: Voices from Beyond the Sexual Binary* (Los Angeles: Alyson Books, 2002), 204–19. Hereafter I cite the version from *GenderQueer*.

23. The term "intersex" has been used interchangeably with and then in place of "hermaphroditism" since the mid-twentieth century to refer to people born with atypical sexual anatomies. Intersex is thus distinct from transgender and transsexuality. Though all these terms are contested, transgender is generally taken to refer to persons who identify as a gender that is different from the one they were assigned at birth, whereas transsexuality is said to denote persons with cross-sex identification who alter their anatomical sex via surgical means. The field of transgender studies has grown exponentially in recent years. For some general points of introduction, see Kate Bornstein, *Gender Outlaw: On Men, Women, and the Rest of Us* (New York: Vintage, 1995); Pat Califia, *Sex Changes: The Politics of Transgenderism* (San Francisco: Cleis Press, 1997); Judith Halberstam, *In a Queer Time and Place: Transgender Bodies, Subcultural Lives* (Durham: Duke University Press, 2005); Paisley Currah, Richard M. Juang, and Shannon Price Minter, "Introduction," in *Transgender Rights*, ed. Paisley Currah, Richard M. Juang, and Shannon Price Minter (Minneapolis: University of Minnesota Press, 2006), xii–xxiv; Susan Stryker, "(De)Subjugated Knowledges: An Introduction to Transgender Studies," in *The Transgender Studies Reader*, ed. Susan Stryker and Stephen Whittle (New York: Routledge, 2006), 1–18; and David Valentine, *Imagining Transgender: An Ethnography of a Category* (Durham: Duke University Press, 2007).

24. Though she would later shift her position (as I detail in chapter 3), Chase's early, politicized understanding of intersex clearly informed her decision to found ISNA in 1993. Chase, "Affronting Reason," 204.

25. Ibid., 205.

26. Ibid., 207.

27. Ibid., 205.

28. Ibid., 211.

29. Ibid., 205.

30. Ibid., 213.

31. Ibid., 211.

32. See Fausto-Sterling, *Sex/Gender*.

33. Butler, *Gender Trouble*.

34. Michel Foucault, *The History of Sexuality, Vol. 1: An Introduction*, trans. Robert Hurley (New York: Vintage, 1990), 140.

35. See Inderpal Grewal, *Transnational America: Feminisms, Diasporas, Neoliberalisms* (Durham: Duke University Press, 2005).

36. This definition is taken from ISNA's website, accessed March 16, 2015, http://www.isna.org/.

37. Thomas Laqueur, *Making Sex: Body and Gender from the Greeks to Freud* (Cambridge: Harvard University Press, 1992); Alice Dreger, *Hermaphrodites and the Medical Invention of Sex* (Cambridge: Harvard University Press, 2000); Elizabeth Reis, *Bodies in Doubt: An American History of Intersex* (Baltimore: Johns Hopkins University Press, 2012); and Lisa Downing, Iain Morland, and Nikki Sullivan, *Fuckology: Critical Essays on John Money's Diagnostic Concepts* (Chicago: University of Chicago Press, 2014).

38. Zine Magubane, "Spectacles and Scholarship: Caster Semenya, Intersex Studies, and the Problem of Race in Feminist Theory," *Signs* 39, no. 3 (2014): 761–85.

39. Karkazis, *Fixing Sex*, 11.

40. On the concept of the naturecultural, see Haraway, *When Species Meet*; and Banu Subramaniam, *Ghost Stories for Darwin: The Science of Variation and the Politics of Diversity* (Urbana: University of Illinois Press, 2014).

41. See Richard Grusin, ed., *The Nonhuman Turn* (Minneapolis: University of Minnesota Press, 2015); Elizabeth Wilson, *Gut Feminism* (Durham: Duke University Press, 2015); Christopher Breu, *Insistence of the Material: Literature in the Age of Biopolitics* (Minneapolis: University of Minnesota Press, 2014); Rosi Bradotti, *The Posthuman* (Cambridge: Polity, 2013); Arthur Kroker, *Body Drift: Butler, Hayles, Haraway* (Minneapolis: University of Minnesota Press, 2012); Mel Chen, *Animacies: Biopolitics, Racial Mattering, and Queer Affect* (Durham: Duke University Press, 2012); Jane Bennett, *Vibrant Matter: A Political Ecology of Things* (Durham: Duke University Press, 2010); Dianna Coole and Samantha Frost, eds., *New Materialisms: Ontology, Agency, and Politics* (Durham: Duke University Press, 2010); Cary Wolfe, *What Is Posthumanism?* (Minneapolis: University of Minnesota Press, 2009); Karen Barad, *Meeting the Universe Halfway: Quantum Physics and the Entanglement of Matter and Meaning* (Durham: Duke University Press, 2007); and Cary Wolfe, ed., *Zoontologies* (Minneapolis: University of Minnesota Press, 2003).

42. Sara Ahmed, "Imaginary Prohibitions: Some Preliminary Remarks on the Founding Gestures of the 'New Materialism,'" *European Journal of Women's Studies* 15, no. 1 (2008): 35.

43. See Cyd Cipolla, Kristina Gupta, David A. Rubin, and Angela Willey, eds., *Queer Feminist Science Studies: A Reader* (Seattle: University of Washington Press, forthcoming).

44. See Vivian May, *Pursuing Intersectionality, Unsettling Dominant Imaginaries* (New York: Routledge, 2015); and Inderpal Grewal and Caren Kaplan, "Transnational Practices and Interdisciplinary Feminist Scholarship: Refiguring Women's and Gender Studies," in *Women's Studies on Its Own*, ed. Robyn Wiegman (Durham: Duke University Press, 2002), 66–81.

45. Kimberle Crenshaw, "Mapping the Margins: Intersectionality, Identity Politics, and Violence against Women of Color," *Stanford Law Review* 43 (1991): 1241–52.

46. The Combahee River Collective, "A Black Feminist Statement," in *Feminist Theory Reader: Local and Global Perspectives*, ed. Carole R. McCann and Seung-kyung Kim (New York: Routledge, 2013), 116–22.

47. Key contributions include Patricia Hill Collins, *Black Feminist Thought: Knowledge, Consciousness, and the Politics of Empowerment* (New York: Routledge, 1990); Collins; *Fighting Words: Black Women and the Search for Justice* (Minneapolis: University of Minnesota Press, 1998); Leslie McCall, "The Complexity of Intersectionality," *Signs* 30, no. 3 (2005): 1771–800; Naomi Zach, *Inclusive Feminism: A Third Wave Theory of Women's Commonality* (Lanham, MD: Rowman and Littlefield, 2005); Jennifer C. Nash, "Re-thinking Intersectionality," *Feminist Review*

89 (2008): 1–15; Robyn Wiegman, *Object Lessons* (Durham: Duke University Press, 2012); Roderick Ferguson, *Aberrations in Black: Toward a Queer of Color Critique* (Minneapolis: University of Minnesota Press, 2003); Chandan Reddy, *Freedom with Violence: Race, Sexuality, and the US State* (Durham: Duke University Press, 2011); and Grace Kyungwon Hong and Roderick Ferguson, eds., *Strange Affinities: The Gender and Sexual Politics of Comparative Racializiation* (Durham: Duke University Press, 2011).

48. Richa Nagar and Amanda Lock Swarr, "Introduction," in *Critical Transnational Feminist Praxis*, ed. Amanda Lock Swarr and Richa Nagar (New York: State University of New York Press, 2010), 5.

49. Inderpal Grewal, *Transnational America: Feminisms, Diasporas, Neoliberalisms* (Durham: Duke University Press, 2005).

50. Karen J. Leong, Roberta Chevrette, Ann Hibner Koblitz, Karen Kuo, and Heather Switzer, "Introduction," *Frontiers: A Journal of Women's Studies* 36, no. 3 (2015): ix.

51. Foucault, *The History of Sexuality, Vol. 1.*

52. Ibid., 154.

53. Butler, *Gender Trouble*, 117.

54. Although some critics have charged Butler with failing to attend to the materiality of sexed bodies, many of these critiques rely upon frameworks that reinstate oppositions between sex and gender, nature and culture, and materiality and discourse—the very binaries Butler's work seeks to disarm and displace. See, for instance, Jordana Rosenberg, "Butler's 'Lesbian Phallus'; or, What Can Deconstruction Feel?" *GLQ* 9, no. 3 (2003): 393–414; Ellen Jean Samuels, "Critical Divides: Judith Butler's Body Theory and the Question of Disability," *NWSA Journal* 14, no. 3 (2002): 58–76; Fiona Webster, "The Politics of Sex and Gender: Benhabib and Butler Debate Subjectivity," *Hypatia* 15, no. 1 (2000): 1–22; Jay Prosser, *Second Skins* (New York: Columbia University Press, 1998); and Pheng Cheah, "Mattering," *Diacritics* 26, no. 1 (1996): 108–39. Butler has responded to some of these critiques in various publications, including *Bodies that Matter* and *Undoing Gender*.

55. Lynne Huffer, *Mad for Foucault: Rethinking the Foundations of Queer Theory* (New York: Columbia University Press, 2010), 47.

56. Ibid., 47.

57. Sally Markowitz, "Pelvic Politics: Sexual Dimorphism and Racial Difference," *Signs* 26, no. 2 (2001): 389–414.

58. Butler, *Gender Trouble*, xxix.

59. See David Rubin, "Women's Studies, Neoliberalism, and the Paradox of the 'Political,'" in *Women's Studies for the Future: Foundations, Interrogations, Politics*, ed. Elizabeth Kennedy and Agatha Beins (New Brunswick: Rutgers University Press, 2005), 245–61.

60. Amrita Basu, "Globalization of the Local/Localization of the Global: Mapping Transnational Women's Movements," *Meridians: Feminism, Race, Transnationalism* 1, no. 1 (2000): 68–84.

61. On the concept of the worlding of the world, see Gayatri Chakravorty Spivak, *In Other Worlds: Essays in Cultural Politics* (New York: Routledge, 1987).

Chapter 1

1. Michel Foucault. "Nietzsche, Genealogy, History," in *The Foucault Reader*, ed. Paul Rabinow (New York: Pantheon Books, 1984), 77.

2. Ibid., 83.

3. Ibid., 83.

4. Inderpal Grewal and Caren Kaplan, eds., *Scattered Hegemonies: Postmodernity and Transnational Feminist Practices* (Minneapolis: University of Minnesota Press, 1994).

5. Robert McRuer, *Crip Theory: Cultural Signs of Queerness and Disability* (New York: New York University Press, 2006).

6. Suzanne J. Kessler, "The Medical Construction of Gender: Case Management of Intersexed Infants," *Signs* 16, no. 1 (1990): 33–38.

7. Ibid., 4.

8. Ibid., 25.

9. Kessler, *Lessons from the Intersexed* (New Brunswick: Rutgers University Press, 1998).

10. See Anne Fausto-Sterling, "The Five Sexes: Why Male and Female Are Not Enough," *Sciences* (March–April, 1993): 20–24; Fausto-Sterling, *Sexing the Body: Gender Politics and the Construction of Sexuality* (New York: Basic Books, 2000); Myra Hird, "Gender's Nature: Intersexuality, Transsexualism and the 'Sex'/'Gender' Binary," *Feminist Theory* 1, no. 3 (2000): 347–63; Sharon Preves, *Intersex and Identity: The Contested Self* (New Brunswick: Rutgers University Press, 2003); and Katrina Karkazis, *Fixing Sex: Intersex, Medical Authority, and Lived Experience* (Durham: Duke University Press, 2008).

11. Fausto-Sterling, *Sexing the Body*; Preves, "Sexing the Intersexed: An Analysis of Sociocultural Responses to Intersexuality," *Signs* 27/2 (2002): 523–56; Karkazis, *Fixing Sex.*

12. Judith Butler, *Gender Trouble: Feminism and the Subversion of Identity* (New York: Routledge, 1990); see also Butler, *Bodies that Matter: On the Discursive Limits of "Sex"* (New York: Routledge, 1993).

13. See the INSA website at http://isna.org.

14. Iain Morland, "Intersex Treatment and the Promise of Trauma," in *Gender and the Science of Difference: Cultural Politics of Contemporary Science and Medicine*, ed. Jill Fisher (New Brunswick: Rutgers University Press, 2011), 156, 157.

15. Elizabeth Reis, *Bodies in Doubt: An American History of Intersex* (Baltimore: Johns Hopkins University Press, 2012).

16. See the ISNA website at http://isna.org.

17. See Fausto-Sterling, *Sexing the Body*; Morgan Holmes, *Intersex: A Perilous Difference* (Selinsgrove: Susquehanna University Press, 2008); and Georgiann

Davis, *Contesting Intersex: The Dubious Diagnosis* (New York: New York University Press, 2015).

18. Jennifer Germon, *Gender: A Genealogy of an Idea* (New York: Palgrave Macmillan, 2009), 1.

19. Bernice Hausman, *Changing Sex: Transsexualism, Technology, and the Idea of Gender* (Durham, Duke University Press, 1995).

20. Lisa Downing, Iain Morland, and Nikki Sullivan, *Fuckology: Critical Essays on John Money's Diagnostic Concepts* (Chicago: The University of Chicago Press, 2015).

21. Sullivan, "The Matter of Gender," in *Fuckology*, 20.

22. Sullivan, "The Matter of Gender, in *Fuckology*, 20.

23. Germon, *Gender*, 3.

24. See Lesley Rogers and Joan Walsh, "Shortcomings of the Psychomedical Research of John Money and Co-Workers into Sex Differences in Behavior: Social and Political Implications," *Sex Roles* 8 (1982): 269–81; Ruth G. Doell and Helen E. Longino, "Sex Hormones and Human Behavior: A Critique of the Linear Model," *Journal of Homosexuality* 15 (1988): 55–78; and Rebecca M. Jordan-Young, *Brain Storm: The Flaws in the Science of Sex Differences* (Cambridge: Harvard University Press, 2010).

25. Ann Oakley, *Sex, Gender and Society* (London: Maurice Temple Smith, 1972), 16.

26. John Hood-Williams, "Goodbye to Sex and Gender," *Sociological Review* 44, no. 1 (1996): 1.

27. Robert Stoller, *Sex and Gender: On the Development of Masculinity and Femininity* (New York: Science House, 1968); and John Money, John Hampson, and Joan Hampson, "Imprinting and the Establishment of Gender Role," *Archives of Neurology and Psychiatry* 77 (1957): 333–36. Scholars such as Germon and Joanne Meyerowitz observe that Stoller himself adopted "gender" from Money's earlier work. See Germon, *Gender*; and Meyerowitz, *How Sex Changed: A History of Transsexuality in the United States* (Cambridge: Harvard University Press, 2002), 114–46.

28. Oakley, *Sex, Gender and Society*, 159.

29. See George Canguilhem, *The Normal and the Pathological*, trans. Carolyn R. Fawcett (1943; repr., Brooklyn: Zone Books, 1991).

30. Rosemarie Garland-Thomson, *Extraordinary Bodies: Figuring Physical Disability in American Culture and Literature* (New York: Columbia University Press, 1997); Sumi Colligan, "Why the Intersexed Shouldn't Be Fixed: Insights from Queer Theory and Disability Studies," in *Gendering Disability*, ed. Bonnie G. Smith and Beth Hutchison (New Brunswick: Rutgers University Press, 2004), 45–60; and McRuer, *Crip Theory*.

31. Oakley, *Sex, Gender and Society*, 160.

32. Ibid., 164.

33. See M. Diamond and H. K. Sigmundson, "Sex Reassignment at Birth: Long-term Review and Implications," *Archives of Pediatric and Adolescent Medicine*

151 (October 1997): 298–304; John Colapinto, *As Nature Made Him: The Boy Who Was Raised a Girl* (New York: HarperCollins, 2000); and Judith Butler, *Undoing Gender* (New York: Routledge, 2004).

34. Vernon A. Rosario, "The History of Aphallia and the Intersexual Challenge to Sex/Gender," in *A Companion to Lesbian, Gay, Bisexual, Transgender, and Queer Studies*, ed. George E. Haggerty and Molly McGarry (London: Blackwell, 2007), 262–81.

35. Oakley, *Sex, Gender and Society*, 189.

36. Ibid.

37. See Denise Riley, *Am I That Name? Feminism and the Category of Women in History* (Minneapolis: University of Minnesota Press, 1988); Dianna Fuss, *Essentially Speaking: Feminism, Nature, and Difference* (New York: Routledge, 1989); Monique Wittig, "One Is Not Born a Woman," in *The Lesbian and Gay Studies Reader*, ed. Henry Abelove, Michele Aina Barale, and David M. Halperin (New York: Routledge, 1993), 103–9; Diane Elam, *Feminism and Deconstruction: Ms. en Abyme* (New York: Routledge, 1994); and Robyn Wiegman, "The Progress of Gender: Whither 'Women'?," in *Women's Studies on Its Own*, ed. Robyn Wiegman (Durham: Duke University Press, 2002), 106–40.

38. Judith Butler, *Gender Trouble*, 53.

39. Ibid., 7.

40. Ibid.

41. Ibid., 141.

42. Ibid.

43. See also Butler, *Bodies that Matter*; and Butler, *Undoing Gender*.

44. Robyn Wiegman, "The Desire for Gender," in *The Blackwell Companion to Lesbian, Gay, Bisexual, Transgender, and Queer Studies*, ed. George E. Haggerty and Molly McGarry (London: Blackwell, 2007), 217–36.

45. Judith Halberstam, *In a Queer Time and Place: Transgender Bodies, Subcultural Lives* (Durham: Duke University Press, 2005); Gayle Salamon, "Transfeminism and the Future of Gender," in *Women's Studies on the Edge*, ed. Joan Wallach Scott (Durham: Duke University Press, 2008), 115–38; and Jean Bobby Noble, *Sons of the Movement: FtMs Risking Incoherence on a Post-Queer Cultural Landscape* (Toronto: Women's Press, 2006).

46. Fausto-Sterling, *Sexing the Body*; Elizabeth A. Wilson, *Neural Geographies: Feminism and the Microstructure of Cognition* (New York: Routledge, 1998); Wilson, *Psychosomatic: Feminism and the Neurological Body* (Durham: Duke University Press, 2004); Wilson, *Gut Feminism* (Durham: Duke University Press, 2015); Karen Barad, *Meeting the Universe Halfway: Quantum Physics and the Entanglement of Matter and Meaning* (Durham: Duke University Press, 2007); Deboleena Roy, "Asking Different Questions: Feminist Practices for the Natural Sciences," *Hypatia* 23, no. 4 (2008): 134–57; and Susan Oyama, Paul E. Griffiths, and Russell D. Grey, eds., *Cycles of Contingency: Developmental Systems and Evolution* (Massachusetts: MIT Press, 2003).

47. Fausto-Sterling, *Sexing the Body*, 4.

48. Ibid., 74.

49. *Oxford English Dictionary* (*OED*), 2nd ed. (1989), s.v. "gender": see 3.b (first use citation 1963).

50. Hausman, *Changing Sex*; and Germon, *Gender*.

51. Hausman, *Changing Sex*, 107.

52. John Money, "Hermaproditism, Gender and Precocity in Hyperadrenocorticism: Psychologic Findings," *Bulletin of the Johns Hopkins Hospital* 96 (1955): 253–64.

53. John Money, "Lexical History and Constructionist Ideology of Gender," in Money, *Gendermaps: Social Constructionism, Feminism, and Sexosophical History* (New York: Continuum, 1995), 18–19.

54. John Money, *Gendermaps: Social Constructionism, Feminism, and Sexosophical History* (New York: Continuum, 1995).

55. See John Money, *Love and Love Sickness: The Science of Sex, Gender Difference, and Pair Bonding* (Baltimore: Johns Hopkins University Press, 1980).

56. Morgan Holmes, *Intersex*, 69.

57. John Money and Anke E. Ehrhardt, *Man & Woman, Boy & Girl: The Differentiation and Dimorphism of Gender Identity from Conception to Maturity* (Baltimore: Johns Hopkins University Press, 1972), 5.

58. Michel Foucault, *Abnormal: Lectures at the Collège de France, 1974–1975*, trans. Graham Burchell (New York: Picador, 2004).

59. Ibid., 68–69.

60. See Chase, "Affronting Reason"; Kessler, *Lessons of the Intersexed*; Fausto-Sterling, *Sexing the Body*; and Holmes, *Intersex*.

61. John Money, "Hermaphroditism: An Inquiry into the Nature of a Human Paradox" (PhD diss., Harvard University, 1952).

62. Ibid., 7.

63. Ibid., 5.

64. Money, "Lexical History," 20.

65. See Meyerowitz, *How Sex Changed*.

66. Money, "Lexical History," 21; emphasis added.

67. Germon, *Gender*, 25.

68. Carole Pateman, *The Sexual Contract* (Stanford: Stanford University Press, 1988).

69. Morland, "Cybernetic Sexology," in *Fuckology*, 101.

70. Hausman, *Changing Sex*, 107.

71. Ibid., 200.

72. Vernon A. Rosario, "Book Review: *Changing Sex: Transsexualism, Technology, and the Idea of Gender*," *Configurations* 4, no. 2 (1996): 243–46.

73. Ibid., 245.

74. Tammassia, cited in Rosario, 244. The original quotation can be found in Arrigio Tamassia, "Sull'inversione dell'instinto sessuale," *Rivista sperimentale di freniatria e di medicina legale* 4 (1878): 99.

75. Morland, "Introduction: Lessons from the Octopus," *GLQ* 15, no. 2 (2009): 191–97.

76. Money, "Hermaphroditism, Gender, and Precocity," 254.

77. Ibid., 258.

78. Gayatri Chakravorty Spivak, *A Critique of Postcolonial Reason: Toward a History of the Vanishing Present* (Cambridge: Harvard University Press, 1999).

79. Ferdinand de Saussure, *Course in General Linguistics*, trans. Roy Harris (1916; repr., Illinois: Open Court Publishing, 1998); see also Barbara Johnson, *A World of Difference* (Baltimore: Johns Hopkins University Press, 1987).

80. Jacques Derrida, *Monolingualism of the Other; or, The Prosthesis of Origin*, trans. Patrick Mensah (Stanford: Stanford University Press, 1998).

81. Downing, Morland, and Sullivan, *Fuckology*, 196.

82. Ibid.

83. See Meyerowitz, *How Sex Changed*.

84. Morland, "Gender, Genitals, and the Meaning of Being Human," in *Fuckology*, 69.

85. Ibid.

86. John Money, Joan G. Hampson, and John L. Hampson, "Imprinting and the Establishment of Gender Role," *Archives of Neurology and Psychiatry* 77 (1957): 333.

87. Rosario, "The History of Aphallia and the Intersexual Challenge to Sex/Gender," 267.

88. Ibid.

89. Alice D. Dreger and April M. Herndon, "Progress and Politics in the Intersex Rights Movement: Feminist Theory in Action," *GLQ* 15, no. 2 (2009): 202.

90. Karkazis, *Fixing Sex*, 7.

91. Holmes, *Intersex*.

92. Hilary Malatino, "Gone, Missing: Queering and Racializing Absence in Trans & Intersex Archives," in *Queer Feminist Science Studies: A Reader*, ed. Cyd Cipolla, Kristina Gupta, David A. Rubin, and Angela Willey (Seattle: University of Washington Press, forthcoming).

93. María Lugones, "Heterosexualism and the Colonial/Modern Gender System," *Hypatia* 22, no. 1 (2007): 195.

94. Malatino, "Gone, Missing."

95. Ibid.

96. Avery Gordon, *Ghostly Matters: Haunting and the Sociological Imaginary* (Minneapolis: University of Minnesota Press, 2008).

97. On hauntology, see Jacques Derrida, *Specters of Marx: The State of the Debt, the Work of Mourning, and the New International*, trans. Peggy Kamuf (New York: Routledge, 1994). On necropolitics, see Achille Mbembe, *On the Postcolony* (Berkeley: University of California Press, 2001).

98. Gayatri Chakravorty Spivak, "Can the Subaltern Speak?," in *Marxism and the Interpretation of Culture*, ed. Cary Nelson and Laurence Grossberg (Urbana: University of Illinois Press, 1988), 271–314.

99. Spivak defines agency against the westocentric humanist grain as "institutionally validated action." See Spivak, *A Critique of Postcolonial Reason*, 71.

100. Malatino, "Gone, Missing."

101. See http://www.intersexualite.org/DSD_warnings.html.

102. Morland, "Postmodern Intersex," in *Ethics and Intersex*, ed. Sharon E. Sytsma (Netherlands: Springer, 2006), 319–31.

103. Butler, *Gender Trouble*, 111.

104. Morland reaches a different conclusion in "Gender, Genitals, and the Meaning of Being Human," in *Fuckology*. He argues that "the challenge for medical reformists . . . is to contest the legacy of Money's treatment model not for its dehumanization of patients, but for its assumptions about their humanity" (90). An earlier version of Morland's essay appeared as "Plastic Man: Intersex, Humanism, and the Reimer Case," *Subject Matters: A Journal of Communications and the Self* 3, nos. 2/4, no. 1 (2007): 81–98.

105. Judith Lorber, *Paradoxes of Gender* (New Haven: Yale University Press, 1995).

Chapter 2

1. Iain Morland, "Postmodern Intersex," in *Ethics and Intersex*, ed. Sharon E. Sytsma (Netherlands: Springer, 2006), 331.

2. For a queer feminist critique of the construction of proper objects, see Judith Butler, "Against Proper Objects," *differences* 6, nos. 2/3 (1994): 1–26.

3. Rachel Lee, "Notes from the (Non)Field: Teaching and Theorizing Women of Color," in *Women's Studies on Its Own*, ed. Robyn Wiegman (Durham: Duke University Press, 2002), 82–105.

4. See Robyn Wiegman, "Academic Feminism Against Itself," *NWSA Journal* 14, no. 2 (2002): 18–34; see also David Rubin, "Women's Studies, Neoliberalism, and the Paradox of the 'Political,'" in *Women's Studies for the Future: Foundations, Interrogations, Politics*, ed. Elizabeth Kennedy and Agatha Beins (New Brunswick: Rutgers University Press, 2005), 245–61.

5. Judith Butler, *Undoing Gender* (New York: Routledge, 2004).

6. Ann Oakley, *Sex, Gender and Society* (London: Maurice Temple Smith, 1972).

7. Gayle Rubin, "The Traffic in Women: Notes on the Political Economy of Sex," in *Toward an Anthropology of Women*, ed. Rayna R. Reiter (Boston: Monthly Review Press, 1975), 157–210.

8. Suzanne J. Kessler, "The Medical Construction of Gender: Case Management of Intersex Infants," *Signs: Journal of Women in Culture and Society* 16, no. 1 (1990): 3–26; *Lessons from the Intersexed* (New Brunswick: Rutgers University Press, 1998).

9. For an updated and expanded version of this aspect of Kessler's argument, see Katrina Karkazis, *Fixing Sex: Intersex, Medical Authority, and Lived Experience* (Durham: Duke University Press, 2008).

10. Suzanne J. Kessler, "The Medical Construction of Gender," 4.

11. See Judith Butler, *Gender Trouble*; Diana Fuss, *Essentially Speaking: Feminism, Nature, and Difference* (New York: Routledge, 1989); Donna Haraway, "The Cyborg Manifesto: Science, Technology, and Socialist-Feminism in the Late Twentieth Century," in *Simians, Cyborgs, and Women: The Reinvention of Nature* (New York: Routledge, 1991), 149–82; and Diane Elam, *Feminism and Deconstruction: Ms. en Abyme* (New York: Routledge, 1994).

12. Gayle Rubin, "The Traffic in Women," 159.

13. Kessler, "The Medical Construction of Gender," 10.

14. Ibid., 4.

15. Ibid., 24.

16. Ibid., 25.

17. Ibid., 24–25.

18. See Michel Foucault, *The Order of Things: An Archeology of the Human Sciences* (New York: Routledge, 1989); Foucault, *The History of Sexuality, Vol. 1: An Introduction*, trans. Robert Hurley (New York: Vintage, 1990); and Foucault, *Abnormal: Lectures at the Collège de France, 1974–1975*, trans. Graham Burchell (New York: Picador, 2004).

19. Ibid., 3 (emphasis added).

20. Kessler, "The Medical Construction of Gender," 25.

21. Kessler, *Lessons from the Intersexed*, 132.

22. Ibid.

23. Anne Fausto-Sterling, "The Five Sexes: Why Male and Female Are Not Enough," *Sciences* 33, no. 2 (1993): 21.

24. Ibid., 20.

25. Ibid., 21.

26. Kessler, *Lessons from the Intersexed*, 90.

27. Fausto-Sterling, "The Five Sexes, Reassessed," *Sciences* 40, no. 4 (2000): 22.

28. Fausto-Sterling, *Sexing the Body: Gender Politics and the Construction of Sexuality* (New York: Basic Books, 2000).

29. Ibid., 5.

30. Ibid., 3.

31. Ibid., 4.

32. This point also appears in Fausto-Sterling's previous book, *Myths of Gender: Biological Theories about Women and Men* (New York: Basic Books, 1985).

33. Fausto-Sterling, *Sexing the Body*, 74.

34. Ibid., 54.

35. Ibid., 79.

36. Ibid., 77.

37. Butler, *Gender Trouble*, 11.

38. Ibid.

39. Ibid., 7.

40. Diane Elam, *Feminism and Deconstruction*, 50.

41. Butler, *Gender Trouble*, 180.

42. Robyn Wiegman, "The Progress of Gender: Whither 'Women'?," in *Women's Studies on Its Own*, ed. Robyn Wiegman (Durham: Duke University Press, 2002), 121.

43. Michel Foucault, *Herculine Barbin: Being the Recently Discovered Memoirs of a Nineteenth-Century French Hermaphrodite*, trans. Richard McDougall (New York: Pantheon, 1980).

44. Butler, *Gender Trouble*, 122.

45. Ibid.

46. Ibid., 179.

47. Michel Foucault, *Herculine Barbin*, xiii.

48. Butler, *Gender Trouble*, 122.

49. Michel Foucault, *Herculine Barbin dite Alexina B.* (Paris: Gallimard, 1978).

50. Michel Foucault, "Le Vrai sexe," *Arcadie* 323 (Novembre 1980): 617–25.

51. Michel Foucault, *Herculine Barbin*, xiii.

52. Butler, *Gender Trouble*, 120.

53. Michel Foucault, *Herculine Barbin*, xiii.

54. "Le Vrai sexe" is reprinted in Foucault's *Dits et Écrits: 1976–1988* (Paris: Gallimard, 2001), 934–42. This quotation appears on 940.

55. I thank Lynne Huffer for translating this passage for me and for sharing and allowing me to draw upon her reading of Foucault's "Introduction."

56. Wiegman, "The Progress of Gender," 120.

57. Ibid., 122.

58. David Valentine and Riki Anne Wilchins, "One Percent on the Burn Chart: Gender, Genitals, and Hermaphrodites with Attitude," *Social Text* 15, nos. 3–4 (1997): 215–22.

59. Valentine and Wilchins, 220.

60. Ibid., 221.

61. Angela Willey, *Undoing Monogamy: The Politics of Science and the Possibilities of Biology* (Durham: Duke University Press, 2016), 159, n. 67. I thank Angie for helping me to think about the problem of intersex exceptionalism.

62. Roderick Ferguson, *Aberrations in Black: Toward a Queer of Color Critique* (Minneapolis: University of Minnesota Press, 2003); Grace Kyungwon Hong, *The Ruptures of American Capital: Women of Color and the Culture of Immigrant Labor* (Minneapolis: University of Minnesota Press, 2006); Grace Hong and Roderick Ferguson, eds., *Strange Affinities: The Gender and Sexual Politics of Comparative Racializiation* (Durham: Duke University Press, 2011); Chandan Reddy, *Freedom with Violence: Race, Sexuality, and the US State* (Durham: Duke University Press, 2011); and Stanley I. Thangaraj, *Desi Hoop Dreams: Pickup Basketball and the Making of Asian-American Masculinity* (New York: New York University Press, 2015).

63. Morgan Holmes, *Intersex: A Perilous Difference* (Selinsgrove: Susquehanna University Press, 2008).

64. Sharon E. Preves, *Intersex and Identity: The Contested Self* (New Brunswick: Rutgers University Press, 2003).

65. Holmes, *Intersex*, 15.

66. Ibid., 15.

67. Ibid., 15.

68. Ibid., 15–16.

69. Ibid., 15.

70. Ibid., 19.

71. See Alice Dreger, ed., *Intersex in the Age of Ethics* (Hagerstown: University Publishing Group, 1999); and Morgan Holmes, ed., *Critical Intersex* (London: Ashgate, 2009).

72. For a representative example of the hypothesis that the destabilization of the subject of women threatens to undo feminism and/or women's studies, see Ellen Messer-Davidow, *Disciplining Feminism*; for a critique of Messer-Davidow's position, see David Rubin, "Women's Studies, Neoliberalism, and the Paradox of the 'Political.'"

73. Jean Bobby Noble, *Sons of the Movement: FtMs Risking Incoherence on a Post-Queer Cultural Landscape* (Toronto: Women's Press, 2006), 42.

74. Gayle Salamon, "Transfeminism and the Future of Gender," in *Women's Studies on the Edge*, ed. Joan W. Scott (Durham: Duke University Press, 2008), 115–38.

75. Ibid., 117.

Chapter 3

1. Morgan Holmes, "Queer Cut Bodies," in *Queer Frontiers: Millennial Geographies, Genders and Generations*, ed. Joseph A Boone et al. (Madison: University of Wisconsin Press, 2000), 102.

2. Cheryl Chase, "Hermaphrodites with Attitude: Mapping the Emergence of Intersex Political Activism," *GLQ* 4, no. 2 (1998): 189–211.

3. Ibid., 201.

4. Suzanne J. Kessler, "The Medical Construction of Gender: Case Management of Intersex Infants," *Signs: Journal of Women in Culture and Society* 16, no. 1 (1990): 3–26; Anne Fausto-Sterling, "The Five Sexes: Why Male and Female Are Not Enough," *Sciences* 33, no. 2 (1993): 20–25; and Cheryl Chase, "Intersexual Rights," *Sciences* 33, no. 4 (1993): 3.

5. Since the turn of the millennium, several major US feminist organizations, including the National Organization for Women, Feminist Majority Foundation, *Ms. Magazine*, and V-Day, have publicly pledged their support for the intersex movement. See "NOW Adopts Intersex Resolution," July 1, 2001, accessed September 18, 2009, http://www.isna.org/node/170; "Medical Violence against Intersex Individuals in the United States," accessed September 18, 2009, http://www.feministcampus.org/fmla/printable-materials/v-day05/intersex_activism.pdf; Martha Coventry, "Making the Cut," *Ms. Magazine*, October–November 2000, accessed September 18, 2009, http://www.msmagazine.com/oct00/makingthecut.html; and "V-Day Endorses ISNA's Mission to End Violence against Intersex People," accessed September 18, 2009, http://www.vday.org/node/1497.html#.WOZWqKIrJTY.

6. Chase, "Hermaphrodites with Attitude," 201.

7. Ibid.

8. Vernon A. Rosario, "An Interview with Cheryl Chase," *Journal of Gay and Lesbian Psychotherapy* 10, no. 2 (2006): 101.

9. Emi Koyama and Lisa Weasel, "From Social Construction to Social Justice: Transforming How We Teach about Intersexuality," *Women's Studies Quarterly* 30, nos. 3–4 (Fall 2002): 169–78.

10. Ibid., 176.

11. My approach in this chapter draws on Sarah Ahmed's methodology in "Imaginary Prohibitions: Some Preliminary Remarks on the Founding Gestures of the 'New Materialism,'" *European Journal of Women's Studies* 15, no. 1 (2008): 23–39.

12. Cheryl Chase, "What Is the Agenda of the Intersex Patient Advocacy Movement?" *Endocrinologist* 13 (2003), 240. This claim is also featured on ISNA's website, accessed December 25, 2015, http://www.isna.org.

13. Gayle Salamon, "Transfeminism and the Future of Gender," in *Women's Studies on the Edge*, ed. Joan W. Scott (Durham: Duke University Press, 2008), 117–116.

14. According to Sharon E. Preves, condition-specific support groups can be placed under the rubric of "intersex advocacy organizations." Such organizations are currently active in "India, Poland, New Zealand, Sweden, Greece, Italy, the Netherlands, Norway, France, Spain, Switzerland, South Africa, Germany, and Australia," as well as in Israel, China, Brazil, Canada, Argentina, and elsewhere. Preves, *Intersex and Identity: The Contested Self* (New Brunswick: Rutgers University Press, 2003), 92.

15. Katrina Karkazis, *Fixing Sex: Intersex, Medical Authority, and Lived Experience* (Durham: Duke University Press, 2008).

16. Ibid., 246.

17. Alice Dreger, *Hermaphrodites and the Medical Invention of Sex* (Cambridge: Harvard University Press, 2000).

18. Susan Stryker, *Transgender History* (London: Seal Press, 2008), 139.

19. Alice Dreger, "Why 'Disorders of Sex Development'? (On Language and Life)," November 17, 2007, accessed March 3, 2014, http://www.alicedreger.com/dsd.html. On the history and politics of the term "queer," see Judith Butler, "Critically Queer," *GLQ* 1, no. 1 (1993): 17–32; and Eve Kosofsky Sedgwick, "Queer and Now," *Tendencies* (Durham: Duke University Press, 1994), 1–22.

20. See David Rubin, "'An Unnamed Blank That Craved a Name': A Genealogy of Intersex as Gender," *Signs* 37, no. 4 (2012): 883–908.

21. See, for instance, Alice Dreger, ed., *Intersex in the Age of Ethics* (Hagerstown: University Publishing Group, 1999); and Georgiann Davis, *Contesting Intersex: The Dubious Diagnosis* (New York: New York University Press, 2015).

22. Susan Stryker, *Transgender History*, 139.

23. Chase, "Hermaphrodites with Attitude," 189.

24. Ibid.

25. Butler, cited in Chase, "Hermaphrodites with Attitude," 208. The original quotation can be found in Judith Butler, *Gender Trouble: Feminism and the Subversion of Identity* (New York: Routledge, 1990), 8.

26. Iain Morland, "The Injured World: Intersex and the Phenomenology of Feeling," *differences* 23, no. 2 (2012): 20–41.

27. Rosario, "An Interview with Cheryl Chase," 104.

28. Ibid., 98.

29. Ibid., 98–99.

30. Michel Foucault, *Abnormal: Lectures at the Collège de France, 1974–1975*, trans. Graham Burchell (New York: Picador, 2004).

31. Rosario, "An Interview with Cheryl Chase," 101.

32. See, among others, Fausto-Sterling, *Sexing the Body*; and Emily Grabham, "Citizen Bodies, Intersex Citizenship," *Sexualities* 10, no. 1 (2007): 29–48.

33. For a Foucauldian critique of the notion of interiority in the context of queer theory, see Lynne Huffer, *Mad for Foucault: Rethinking the Foundations of Queer Theory* (New York: Columbia University Press, 2010).

34. See Morland, "Gender, Genitals, and the Meaning of Being Human," in *Fuckology: Critical Essays on John Money's Diagnostic Concepts*, ed. Lisa Downing, Iain Morland, and Nikki Sullivan (Chicago: University of Chicago Press, 2015), 69–98.

35. Alice D. Dreger and April M. Herndon, "Progress and Politics in the Intersex Rights Movement: Feminist Theory in Action," *GLQ* 15, no. 2 (2009): 199–224.

36. Dreger and Herndon, "Progress and Politics in the Intersex Rights Movement," 216. The original citation at the end of this passage can be found in Cheryl Chase, "What Is the Agenda of the Intersex Patient Advocacy Movement?," *Endocrinologist* 13 (2003): 240.

37. April Herndon, "Why Doesn't ISNA Want to Eradicate Gender?," February 17, 2006, accessed February 18, 2009, http://www.isna.org/faq/not_eradicating_gender.

38. For a comparative analysis of single-issue versus multi-issue organizing, see Urvashi Vaid, *Virtual Equality: The Mainstreaming of Lesbian and Gay Liberation* (New York: Anchor Books, 1996).

39. See Nancy C. M. Hartsock, "The Feminist Standpoint: Developing the Ground for a Specifically Feminist Historical Materialism," in *Discovering Reality: Feminist Perspectives on Epistemology, Metaphysics, Methodology, and Philosophy of Science*, ed. Sandra Harding and Merrill B. Hintikka (Dordrecht, Holland: Springer, 1983), 283–310; Patricia Hill Collins, *Black Feminist Thought: Knowledge, Consciousness, and the Politics of Empowerment* (New York: Routledge, 1990); and Donna Haraway, "Situated Knowledges: The Science Question in Feminism and the Privilege of Partial Perspective," in *Simians, Cyborgs, and Women: The Reinvention of Nature* (New York: Routledge, 1991), 183–202.

40. Patricia Hill Collins, "Defining Black Feminist Thought," in *Feminist Theory Reader: Local and Global Perspectives*, ed. Carole R. McCann and Seung-kyung Kim (New York: Routledge, 2013), 385.

41. Joan W. Scott, "Experience," in *The Lesbian and Gay Studies Reader*, ed. Henry Abelove, Michele Aina Barale, and David M. Halperin (New York: Routledge, 1993), 397–415.

42. Ibid., 399–400.

43. Koyama and Weasel, "From Social Construction to Social Justice," 176.

44. Ibid., 170.

45. See, among many others, Trinh T. Minh-ha, *Woman, Native, Other: Writing Postcoloniality and Feminism* (Bloomington: Indiana University Press, 1989); bell hooks, *Feminist Theory: From Margin to Center* (Boston: South End Press, 1984); Linda Alcoff, "The Problem of Speaking for Others," in *Who Can Speak? Authority and Critical Identity*, ed. Judith Roof and Robyn Wiegman (Urbana: University of Illinois Press, 1995), 97–119; and Norma Alarcón, "The Theoretical Subject(s) of *This Bridge Called My Back* and Anglo-American Feminism," in *Haciendo Caras: Making Face, Making Soul: Creative and Critical Perspectives by Feminists of Color*, ed. Gloria Anzaldua (San Francisco: Aunt Lute Books, 1990), 356–69.

46. See Eve Kosofsky Sedgwick, *Between Men: English Literature and Male Homosocial Desire* (New York: Columbia University Press, 1986); Monique Wittig, "One Is Not Born a Woman," in *The Lesbian and Gay Studies Reader*, ed. Henry Abelove, Michele Aina Barale, and David M. Halperin (New York: Routledge, 1993), 103–9; Butler, *Gender Trouble*; and Judith Halberstam, *Female Masculinity* (Durham: Duke University Press, 1998).

47. Lila Abu-Lughod, "Do Muslim Women Really Need Saving? Anthropological Reflections on Cultural Relativism and Its Others," *American Anthropologist* 104, no. 3 (2002): 783–90.

48. Vernon A. Rosario, "The History of Aphallia and the Intersexual Challenge to Sex/Gender," in *A Companion to Lesbian, Gay, Bisexual, Transgender, and Queer Studies*, ed. George E. Haggerty and Molly McGarry (London: Blackwell, 2007), 262–81.

49. Ibid., 274.

50. Vernon A. Rosario, "The Biology of Gender and the Construction of Sex?," *GLQ* 10, no. 2 (2004): 280–87.

51. Rosario, "The Biology of Gender," 282.

52. Ibid., 283–84.

53. Ibid., 284.

54. Holmes, *Intersex: A Perilous Difference* (Selinsgrove: Susquehanna University Press, 2008), 23.

55. Rosario, "The History of Aphallia," 275.

56. Ibid., 275.

57. Ibid., 276.

58. In using "deconstruction," which is actually a technical term, in this way, these authors participate in a more general trend in the humanities and social sciences, which Jean-Michel Rabate has analyzed as a backlash against poststructuralism. See Jean-Michel Rabate, *The Future of Theory* (London: Blackwell, 2002).

59. Deconstruction names the instability of meaning that arises through the structure of the supplement, trace, or différance, the continuous process of difference-as-deferral through which binary, symmetrical meanings are shown to be asymmetrical. See, among many other texts, Jacques Derrida, *Of Grammatology*, trans. Gayatri Chakravorty Spivak (Baltimore: Johns Hopkins University Press, 1976).

60. Koyama and Weasel's and Rosario's frequent, derogatory usage of the term "deconstruction" might be read to suggest the degree to which they have registered and object to the impact of Judith Butler's work on thinking about intersex issues. While Koyama and Weasel do not cite Butler, Rosario does. In "The Biology of Gender and the Construction of Sex?," Rosario cites Butler's "Doing Justice to Someone" as he laments the ways in which, using the David Reimer case, "feminist and queer academics [have] turned intersex into the next great hope for deconstructing sex/gender" (283). From this perspective, Rosario's critique of "deconstruction," and perhaps Koyama and Weasel's as well, can be read as being primarily a critique of Butler, even though she is not named as the primary object of their critiques. If that is indeed the case, Koyama and Weasel and Rosario are surely not alone in their dismissal of Butler's interventions. See, for instance, Martha Nussbaum, "The Professor of Parody," *The New Republic* 220, no. 16 (February 22, 1999): 37–45. For a compelling critique of Nussbaum's position, see Robyn Wiegman, "Feminism, Institutionalism, and the Idiom of Failure," *differences* 11, no. 3 (1999): 107–36.

61. Karkazis, *Fixing Sex*, 247.

62. Ibid., 14.

63. See ISNA's website, accessed December 25, 2015, http://www.isna.org.

64. In addition, in reiterating ISNA's position, Karkazis also explicitly validates ISNA's view that "most feminist interest in intersexuality" is concerned "to deconstruct or eliminate gender, or to advocate for a third sex or no sex." As I suggested above, and as I show below in greater detail, this claim is arguably an inaccurate characterization of feminist studies of intersex.

65. Butler, cited in Noah Ben-Asher, "The Necessity of Sex Change: A Struggle for Transsex and Intersex Liberties," *Harvard Journal of Law and Gender* 29, no. 1 (2006): 71. Original in Butler, *Undoing Gender* (New York: Routledge, 2004), 7–8.

66. Ben-Asher, "The Necessity of Sex Change," 72.

67. Ibid.

68. See Michel Foucault, *The History of Sexuality, Volume II: The Use of Pleasure*, trans. Robert Hurley (New York: Vintage Books, 1990); and Luther H. Martin, Huck Gutman, and Patrick H. Hutton, eds., *Technologies of the Self: A Seminar with Michel Foucault* (Massachusetts: University of Massachusetts Press, 1988). For a cogent analysis of Foucault's work on these topics, see Lynne Huffer, *Mad for Foucault*.

69. See Butler, *Gender Trouble*; Sandy Stone, "The Empire Strikes Back: A Posttransexual Manifesto," in *The Transgender Studies Reader*, ed. Susan Stryker and Stephen Whittle (New York: Routledge, 2006), 221–35; Naomi Wolf, *The Beauty Myth: How Images of Beauty Are Used against Women* (New York: Harper,

2004); Patricia Hill Collins, *Fighting Words: Black Women and the Search for Justice* (Minneapolis: University of Minnesota Press, 1998); Michael Kimmel, *Guyland: The Perilous World Where Boys Become Men* (New York: Harper, 2006); Judith Halberstam, *Female Masculinity* (Durham: Duke University Press, 1998); and Roderick Ferguson, *Aberrations in Black: Toward a Queer of Color Critique* (Minneapolis: University of Minnesota Press, 2003).

70. Kate Bornstein, *Gender Outlaw: On Men, Women, and the Rest of Us* (New York: Vintage, 1995).

71. Alice Dreger, *Hermaphrodites and the Medical Invention of Sex* (Cambridge: Harvard University Press, 1998), 8.

72. Ibid., 9.

73. Gayatri Chakravorty Spivak, *The Postcolonial Critic: Interviews, Strategies, Dialogues*, ed. Sarah Harasym (New York: Routledge, 1990), 122.

74. See Cyd Cipolla, Kristina Gupta, David A. Rubin, and Angela Willey, eds., *Queer Feminist Science Studies: A Reader* (Seattle: University of Washington Press, forthcoming).

75. See Lee Edelman, *No Future: Queer Theory and the Death Drive* (Durham: Duke University Press, 2004); and Elizabeth Wilson, *Gut Feminism* (Durham: Duke University Press, 2015).

76. Wilson, *Gut Feminism*, 6.

77. See David Rubin, "Women's Studies, Neoliberalism, and the Paradox of the 'Political,'" in *Women's Studies for the Future: Foundations, Interrogations, Politics*, ed. Elizabeth Kennedy and Agatha Beins (New Brunswick: Rutgers University Press, 2005), 245–61.

78. Ben-Asher, "The Necessity of Sex Change," 75.

Chapter 4

1. Alice Dreger, "Intersex and Human Rights: The Long View," in *Ethics and Intersex*, ed. Sharon E. Sytsma (Netherlands: Springer, 2006), 79.

2. Elizabeth Reis, *Bodies in Doubt: An American History of Intersex* (Baltimore: Johns Hopkins University Press, 2012).

3. In pursuing this line of argument, I build on Iain Morland's insight that intersex activism deserves just as careful critical scrutiny as does medical dogma around intersex. Iain Morland, "Between Critique and Reform: Ways of Reading the Intersex Controversy," in *Critical Intersex*, ed. Morgan Holmes (Farnham: Ashgate, 2009), 191.

4. Leela Fernandes, *Transnational Feminism in the United States: Knowledge, Ethics, Power* (New York: New York University Press, 2013).

5. Richa Nagar and Amanda Lock Swarr, "Introduction," in *Critical Transnational Feminist Praxis*, ed. Amanda Lock Swarr and Richa Nagar (New York: State University of New York Press, 2010), 5.

6. Sandra K. Soto, *Reading Chican@ Like a Queer: The Demastery of Desire* (Austin: University of Texas Press, 2011), 1.

7. Tani E. Barlow, "'green blade in the act of being grazed': Late Capital, Flexible Bodies, Critical Intelligibility," *differences: A Journal of Feminist Cultural Studies* 10, no. 3 (1998): 119.

8. I use the term "sex/gender" in this chapter to emphasize the dynamic co-constitutive relationship between social and biological systems in the formation of human embodiment and subjectivity. See Anne Fausto-Sterling, *Sex/Gender: Biology in a Social World* (New York: Routledge, 2012).

9. Laura Briggs, Gladys McKormick, and J. T. Way, "Transnationalism: A Category of Analysis," *American Quarterly* 60, no. 3 (2008): 625–48.

10. Caren Kaplan and Inderpal Grewal, "Transnational Practices and Interdisciplinary Feminist Scholarship: Refiguring Women's and Gender Studies," in *Women's Studies on Its Own*, ed. Robyn Wiegman (Durham: Duke University Press, 2002), 73.

11. Caren Kaplan, *Questions of Travel: Postmodern Discourses of Displacement* (Durham: Duke University Press, 1996), 169.

12. Emma Perez, *The Decolonial Imaginary: Writing Chicanas into History* (Bloomington: Indiana University Press, 1999).

13. Susan Stryker and Paisley Currah, "General Editors Introduction," *Transgender Studies Quarterly* 1, no. 3 (2014): 303–4.

14. Dipesh Chakrabarty, *Provincializing Europe: Postcolonial Thought and Historical Difference* (Princeton: Princeton University Press, 2008).

15. ISNA Press Release, "Intersex Declared a Human Rights Issue," May 5, 2005, accessed March 2, 2014, http://www.isna.org/node/841. The press release does not contain page numbers.

16. Ibid.

17. Ibid.

18. Julie A. Greenberg, *Intersexuality and the Law: Why Sex Matters* (New York: New York University Press, 2012).

19. ISNA Press Release, "ISNA Honored with Human Rights Award," 2000, accessed March 9, 2014, http://www.isna.org/node/15.

20. Marcus de Maria Arana (principal author), "A Human Rights Investigation into the Medical 'Normalization' of Intersex People: A Report of a Public Hearing by the Human Rights Commission of the City and County of San Francisco," April 28, 2005, accessed March 9, 2014, http://www.isna.org/files/SFHRC_Intersex_Report.pdf; and European Network of Legal Experts, "Trans and Intersex People: Discrimination on the Grounds of Sex, Gender Identity and Gender Expression," Luxembourg: European Union, 2011, accessed December 1, 2014, http://www.coe.int/t/dg4/lgbt/Source/trans_and_intersex_people_EC_EN.pdf.

21. Alice Dreger, "Ending Forced 'Genital-Normalizing' Surgeries," *The Atlantic*, February 25, 2013, accessed March 9, 2014, http://www.theatlantic.com/health/archive/2013/02/ending-forced-genital-normalizing-surgeries/273300/; and

Organization of American States, "Rights of Lesbian, Gay, Bisexual, Trans, and Intersex Persons," 2013, accessed July 14, 2014, http://www.oas.org/en/iachr/lgtbi/.

22. The full text of the Australian Parliament's "Sex Discrimination Amendment (Sexual Orientation, Gender Identity and Intersex Status) Bill, accessed March 9, 2014, http://www.aph.gov.au/Parliamentary_Business/Bills_Legislation/Bills_Search_Results/Result?bId=r5026. The new German law is described in Jacinta Nandl, "Germany Got It Right by Offering a Third Gender Option on Birth Certificates," *The Guardian*, November 10, 2013, accessed March 9, 2014, http://www.theguardian.com/commentisfree/2013/nov/10/germany-third-gender-birth-certificate. German intersex activists have critiqued the legislation on gender classification, arguing that the new law does not demedicalize intersex and consequently does not address the human rights violations intersex people are subject to. See Silvan Agius, Morgan Carpenter, and Dan Christian Ghattas, "Third Gender: A Step Toward Ending Intersex Discrimination," *Spiegel Online International*, August 22, 2013, accessed December 1, 2014, http://www.spiegel.de/international/europe/third-gender-option-in-germany-a-small-step-for-intersex-recognition-a-917650.html.

23. Dan Christian Ghattas, *Human Rights Between the Sexes: A Preliminary Study on the Life Situations of Inter* Individuals* (Berlin: Heinrich-Boll-Stiftung, 2013).

24. Meredith Bennett-Smith, "Mark and Pam Crawford, Parents of Intersex Child, Sue South Carolina for Sex Assignment Surgery," *The Huffington Post*, May 5, 2013, accessed March 9, 2014, http://www.huffingtonpost.com/2013/05/15/mark-pam-crawford-intersex-child_n_3280353.html.

25. Inderpal Grewal, *Transnational America: Feminisms, Diasporas, Neoliberalisms* (Durham: Duke University Press, 2005), 121.

26. Lila Abu-Lughod, "Do Muslim Women Really Need Saving? Anthropological Reflections on Cultural Relativism and Its Others," *American Anthropologist* 104, no. 3 (2002): 783–90.

27. Wendy Brown, "Suffering the Paradoxes of Rights," in *Left Legalism/Left Critique*, ed. Wendy Brown and Janet Halley (Durham: Duke University Press, 2002), 422.

28. Michel Foucault, *The History of Sexuality, Vol. 1: An Introduction* (New York: Vintage, 1990).

29. See, among many others, Lauren Berlant, *The Queen of America Goes to Washington City: Essays on Sex and Citizenship* (Durham: Duke University Press, 1997); Barbara Cruikshank, *The Will to Empower: Democratic Citizens and Other Subjects* (Ithaca: Cornell University Press, 1999); Dean Spade, *Normal Life: Administrative Violence, Critical Trans Politics, and the Limits of the Law* (Boston: South End Press, 2011); and Stanley I. Thangaraj, *Desi Hoop Dreams: Pickup Basketball and the Making of Asian-American Masculinity* (New York: New York University Press, 2015).

30. Brown, "Suffering the Paradoxes of Rights," 422.

31. Pheng Cheah, *Inhuman Conditions: On Cosmopolitanism and Human Rights* (Cambridge: Harvard University Press, 2007).

32. See David Harvey, *A Brief History of Neoliberalism* (Oxford: Oxford University Press, 2005); Aihwa Ong, *Neoliberalism as Exception: Mutations in Citizen-*

ship and Sovereignty (Durham: Duke University Press, 2006); and Miranda Joseph and David Rubin, "Promising Complicities: On the Sex, Race, and Globalization Project," in *A Companion to Lesbian, Gay, Bisexual, Transgender, and Queer Studies*, ed. George E. Haggerty and Molly McGarry (London: Blackwell, 2007), 430–51.

33. Lisa Duggan, *The Twilight of Equality? Neoliberalism, Cultural Politics, and the Attack on Democracy* (Boston: Beacon Press, 2004).

34. Grace Kyungwon Hong, *The Ruptures of American Capital: Women of Color Feminism and the Culture of Immigrant Labor* (Minneapolis: University of Minnesota Press, 2006); and Chandan Reddy, *Freedom with Violence: Race, Sexuality, and the US State* (Durham: Duke University Press, 2011).

35. Iris Marion Young, *Global Challenges: War, Self-Determination, and Responsibility for Justice* (New York: Polity, 2006).

36. Jasbir K. Puar, *Terrorist Assemblages: Homonationalism in Queer Times* (Durham: Duke University Press, 2007).

37. De-medicalization remains a central goal of a number of contemporary intersex alliances, including Intersex South Africa (founded in 2000 by the late Sally Gross), Bodies Like Ours (founded in 2002 by Peter Trinkle), and OII (founded in 2003 by Curtis Hinkle).

38. Noah Ben-Asher, "The Necessity of Sex Change: A Struggle for Transsex and Intersex Liberties," *Harvard Journal of Law and Gender* 29, no. 1 (2006): 62.

39. Ibid., 73.

40. Cheryl Chase, "Hermaphrodites with Attitude: Mapping the Emergence of Intersex Political Activism" *GLQ* 4, no. 2 (1998): 204.

41. Cheryl Chase, "'Cultural Practice' or 'Reconstructive Surgery'? US Genital Cutting, the Intersex Movement, and Medical Double Standards," in *Genital Cutting and Transnational Sisterhood: Disputing US Polemics*, ed. Stanlie M. James and Claire C. Roberston (Urbana: University of Illinois Press, 2002), 144–45.

42. Miranda Joseph, "Analogy and Complicity: Women's Studies, Lesbian/Gay Studies, and Capitalism," in *Women's Studies on Its Own*, ed. Robyn Wiegman (Durham: Duke University Press, 2002), 267–92.

43. Ben Asher, "The Necessity of Sex Change," 73.

44. See, for instance, Robin Morgan and Gloria Steinem, "The International Crime of Genital Mutilation," *Ms.* (March 1980): 65–67; Mary Daly, *Gyn/Ecology: The Metaethics of Radical Feminism* (Boston: Beacon Press, 1978); and Alice Walker, *Possessing the Secret of Joy* (New York: Pocket Books, 1992).

45. Leslye Obiora, "Bridges and Barricades: Rethinking Polemics and Intransigence in the Campaign Against Female Circumcision," in *Global Critical Race Feminism: An International Reader*, ed. Adrien Katherine Wing (New York: New York University Press, 2000), 260–74; and Rogaia Abusharaf, "Unmasking Tradition," *Sciences* (March/April 1998): 23–27.

46. Wairimu Ngaruiya Njambi, "Dualisms and Female Bodies in Representations of African Female Circumcision: A Feminist Critique," *Feminist Theory* 5, no. 3 (2004): 281–303.

47. Njambi, "Dualisms and Female Bodies," 281.

48. Ibid., 295.

49. Ibid., 299.

50. See Abu-Lughod, "Do Muslim Women Really Need Saving?"; Saba Mahmood, *Politics of Piety: The Islamic Revival and the Feminist Subject* (Princeton: Princeton University Press, 2005); and Gayatri Chakravorty Spivak, *A Critique of Postcolonial Reason: Toward a History of the Vanishing Present* (Cambridge: Harvard University Press, 1999).

51. Claire Hemmings, *Why Stories Matter: The Political Grammar of Feminist Theory* (Durham: Duke University Press, 2011), 224.

52. Claire C. Robertson, "Getting Beyond the Ew! Factor: Rethinking US Approaches to African Genital Cutting," in *Genital Cutting and Transnational Sisterhood: Disputing US Polemics*, ed. Stanlie M. James and Clair C. Robertson (Urbana: University of Illinois Press, 2002), 54–86.

53. Nikki Sullivan, "'The Price to Pay for Our Common Good': Genital Modification and the Somatechnologies of Cultural (In)Difference," *Social Semiotics* 17, no. 3 (2007): 395–409.

54. Ben-Asher, "The Necessity of Sex Change," 73.

55. Abusharaf, "Unmasking Tradition," 26–27.

56. Njambi, "Dualisms and Female Bodies," 299.

57. Ben-Asher, "The Necessity of Sex Change," 75.

58. Cited in Ben-Asher, "The Necessity of Sex Change," 73. Original in 18 U.S.C. § 116 (Supp. III 1997).

59. Cheryl Mattingly and Linda C. Garro, eds., *Narrative and the Cultural Construction of Illness and Healing* (Berkeley: University of California Press, 2000).

60. See Susan Bordo, *Unbearable Weight: Feminism, Western Culture, and the Body* (Berkeley: University of California Press, 2004).

61. ISNA Press Release, "Colombia High Court Restricts Surgery on Intersex Children," 1999, accessed March 9, 2014, http://www.isna.org/colombia/.

62. Julie A. Greenberg and Cheryl Chase, "Background of the Colombia Decisions," 1999, accessed March 9, 2014, http://www.isna.org/node/21. Much has been written about the life of David Reimer. Reimer was treated by John Money (the principal architect of the western biomedical paradigm of intersex management), and Money infamously used this case to justify his theory that surgical and hormonal intervention could guarantee normative gender-role/identity. However, critics cite the Reimer story to call attention to the bioethical abuses of the medical normalization of intersex bodies. For a comprehensive analysis, see Iain Morland, "Gender, Genitals, and the Meaning of Being Human," in *Fuckology: Critical Essays on John Money's Diagnostic Concepts*, ed. Lisa Downing, Iain Morland, and Nikki Sullivan (Chicago: University of Chicago Press, 2015), 69–98.

63. Ben-Asher, "The Necessity of Sex Change," 66, n. 76. A partial translation of the court's decision in English can be found in "The Rights of Intersexed Infants and Children: Decision of the Colombian Constitutional Court, Bogota, Colombia, May 12, 1999 (SU-337/99)," trans. Nohemy Solozano-Thompson, in *Transgender Rights*, ed. Paisley Currah, Richard M. Juang, and Shannon Price Minter (Minneapolis: University of Minnesota Press, 2006), 122–40.

64. According to Greenberg and Chase, "among sources consulted domestically were the Medical Faculty of the University of Javeriana, the Colombian Psychological Society, the Colombian Psychiatric Society, the Colombian Society for Urology, and Dr. Bernard Ochoa (a surgeon who is considered Colombia's foremost scientific authority on intersexuality)." Greenberg and Chase, "Background of the Colombia Decisions," http://www.isna.org/node/21.

65. "The Rights of Intersexed Infants and Children," 136–37.

66. Greg Grandin, *Empire's Workshop: Latin America, The United States, and the Rise of the New Imperialism* (New York: Holt Paperbacks, 2006).

67. Vek Lewis, *Crossing Sex and Gender in Latin America* (New York: Palgrave MacMillan, 2010), 5.

68. See Don Kulick, *Travesti: Sex, Gender, and Culture among Brazilian Transgendered Prostitutes* (Chicago: University of Chicago Press, 1998); Pérez, *The Decolonial Imaginary*; and Ángela Ixkic Bastian Duarte, "From the Margins of Latin American Feminism: Indigenous and Lesbian Feminisms," *Signs: Journal of Women in Culture and Society* 38, no. 1 (2012): 153–78.

69. Anibal Quijano, "Coloniality of Power, Eurocentrism, and Latin America," *Nepantla: Views from South* 1, no. 3 (2000): 533–80.

70. See Sander L. Gilman, *Difference and Pathology: Stereotypes of Sexuality, Race, and Madness* (Ithaca: Cornell University Press, 1985); and Stephen Jay Gould, *The Mismeasure of Man* (New York: W.W. Norton, 1996).

71. Hilary Malatino, "Situating Bio-Logic, Refiguring Sex: Intersexuality and Coloniality," in *Critical Intersex*, ed. Morgan Holmes (Farnham: Ashgate, 2009), 73–96.

72. Zine Magubane, "Spectacles and Scholarship: Caster Semenya, Intersex Studies, and the Problem of Race in Feminist Theory," *Signs* 39, no. 3 (2014): 761–85.

73. María Lugones, "Heterosexualism and the Colonial/Modern Gender System," *Hypatia* 22, no. 1 (2007): 195.

74. Ibid., 195–96.

75. See ISNA's website, accessed July 31, 2015, http://www.isna.org/.

76. Foucault, *The History of Sexuality, Volume One*, 140. On biopolitics, see also Ann Laura Stoler, *Race and the Education of Desire: Foucault's History of Sexuality and the Colonial Order of Things* (Durham: Duke University Press, 1995); and Achile Mbembe, *On the Postcolony* (Berkeley: University of California Press, 2001).

77. "The Rights of Intersexed Infants and Children," 125.

78. Page numbers are not provided in "ISNA's Amicus Brief on Intersex Genital Surgery," and therefore I cite the brief in what follows without page numbers.

79. Ben-Asher, "The Necessity of Sex Change," 75.

80. Starting in 2007, various reports began to circulate in the Colombian media about the practice of FGC by the indigenous Embera Chami people, an aboriginal group who live in various rural regions, mostly in the province of Risaralda. Prior to this date, there is no formal record of anti-FGM activism in Colombia. See Angela Castellanos, "Colombia Confronts Female Genital Mutilation," *HR Reality Check*, August 18, 2008, accessed July 14, 2014, http://rhrealitycheck.org/article/2008/08/18/

colombia-confronts-female-genital-mutilation/; and Scott Kobewka, "Embera Vow to Stop Female Genital Mutilation," *Colombia Reports*, November 23, 2010, accessed July 15, 2014, http://colombiareports.co/embera-vow-stop-female-genital-mutilation/.

81. See Constitutional Court of Colombia, Sentencia No. T-477/95 (1999), 5.2.1.

82. "The Rights of Intersexed Infants and Children," 131.

83. Ibid., 135.

84. Patricia González Sánchez, Catalina Velásquez Acevedo, and Sandra Patricia Duque Quintero, "Problemática Jurídica de Los Estados Intersexuales: El Caso Colombiano," ["Legal Problems of Intersex States: The Colombian Case,"] *Iatriea* 23, no. 3 (2010): 204–11.

85. Morgan Holmes, "Deciding Fate or Protecting a Developing Autonomy? Intersex Children and the Colombian Constitutional Court," in *Transgender Rights*, ed. Paisley Currah, Richard M. Juang, and Shannon Price Minter (Minneapolis: University of Minnesota Press, 2006), 102–21.

86. Ibid., 112.

87. Ibid., 113.

88. Ibid., 117.

89. Ibid., 118–19.

90. Julia Sandra Bernal Crespo, "Estados Intersexuales en Menores de Edad: Los Principios de Autonomía y Beneficencia," ["Intersex States in Minors: The Principles of Autonomy and Beneficence,"] *Revista de Derecho* 36 (2011): 53–86.

91. Ibid., 82 (author's translation).

92. See also Anne Tamar-Mattis and Milton Diamond, "Managing Variations in Sex Development," *Journal of Pediatric Endocrinology and Metabolism* 20, no. 4 (2007): 552–53; and Elizabeth Reis, "Divergence or Disorder? The Politics of Naming Intersex," *Perspectives in Biology and Medicine* 50, no. 4 (2007): 535–43.

93. Stephen Kerry, "'Intersex Imperialism' and the Case of Caster Semenya: The Unacceptable Woman's Body," *Scan: Journal of Media, Arts, Culture* 8, no. 1 (2011), accessed July 1, 2015, http://scan.net.au/scan/journal/display.php?journal_id=158.

94. Mauro Cabral, "The Marks on Our Bodies," *Intersex Day*, October 25, 2015, http://intersexday.org/en/mauro-cabral-marks-bodies/.

95. See Chela Sandoval, *Methodology of the Oppressed* (Minneapolis: University of Minnesota Press, 2000).

96. Chakrabarty, *Provincializing Europe*, xiii.

Chapter 5

1. Sally Markowitz, "Pelvic Politics: Sexual Dimorphism and Racial Difference," *Signs* 26, no. 2 (2001): 389–414.

2. Havelock Ellis, *Studies in the Psychology of Sex*, 2 vols. (New York: Random House, 1905); see also Ellis, *Sexual Inversion* (London: University Press, 1897).

3. Peter Hegarty, *Gentlemen's Disagreement: Alfred Kinsey, Lewis Terman, and the Sexual Politics of Smart Men* (Chicago: University of Chicago Press, 2013).

4. Markowitz, "Pelvic Politics," 391.

5. As I argued in chapter 1, *gender* did not enter into western medico-scientific, theoretical, and sociopolitical discourses until Money made his decisive contribution to the mid-twentieth century medicalization of intersex bodies.

6. Markowitz, "Pelvic Politics," 390.

7. Ibid., 410.

8. See Vivian May, *Pursuing Intersectionality, Unsettling Dominant Imaginaries* (New York: Routledge, 2015).

9. The literature on intersectionalty is gargantuan. Key works on the topic include Kimberle Crenshaw, "Demarginalizing the Intersection of Race and Sex: A Black Feminist Critique of Antidiscrimination Doctrine, Feminist Theory and Antiracist Politics," *University of Chicago Legal Forum* 140 (1989): 139–67; Crenshaw, "Mapping the Margins: Intersectionality, Identity Politics, and Violence against Women of Color," *Stanford Law Review* 43 (1991): 1241–52; Patricia Hill Collins, *Black Feminist Thought: Knowledge, Consciousness, and the Politics of Empowerment* (New York: Routledge, 1990); Collins; *Fighting Words: Black Women and the Search for Justice* (Minneapolis: University of Minnesota Press, 1998); Leslie McCall, "The Complexity of Intersectionality," *Signs* 30, no. 3 (2005): 1771–800; Naomi Zach, *Inclusive Feminism: A Third Wave Theory of Women's Commonality* (Lanham, MD: Rowman and Littlefield, 2005); Jennifer C. Nash, "Re-thinking Intersectionality," *Feminist Review* 89 (2008): 1–15; Robyn Wiegman, *Object Lessons* (Durham: Duke University Press, 2012); and May, *Pursuing Intersectionality, Unsettling Dominant Imaginaries.*

10. Markowitz, "Pelvic Politics," 411.

11. Gilbert Herdt, ed., *Third Sex, Third Gender: Beyond Sexual Dimorphism in History and Culture* (London: Zone Books, 1996).

12. Markowitz, "Pelvic Politics," 391.

13. Ibid.

14. Ibid., 411.

15. See Zine Magubane, "Spectacles and Scholarship: Caster Semenya, Intersex Studies, and the Problem of Race in Feminist Theory," *Signs* 39, no. 3 (2014): 761–85.

16. Jennifer Doyle, "Dirt Off Her Shoulders," *GLQ* 19, no. 4 (2013): 421.

17. Gayatri Chakravorty Spivak, "Can the Subaltern Speak?," in *Marxism and the Interpretation of Culture*, ed. Cary Nelson and Laurence Grossberg (Urbana: University of Illinois Press, 1988), 271–314.

18. Christopher Clarey, "Gender Test after Gold Medal Finish," *New York Times*, August 19, 2009, http://www.nytimes.com/2009/08/20/sports/20runner.html.

19. Doyle, "Dirt Off Her Shoulders," 420.

20. Judith Butler, *Undoing Gender* (New York: Routledge, 2004), 57–76.

21. Doyle, "Dirt Off Her Shoulders," 420. For a feminist analysis of how biology and culture shape bodily capacities and practices of comportment in gendered ways, see Fausto-Sterling, *Sexing the Body: Gender Politics and the Construction of Sexuality* (New York: Basic Books, 2000).

22. Doyle, "Dirt Off Her Shoulders," 420.

23. Fausto-Sterling, "The Five Sexes, Reassessed," *Sciences* 40, no. 4 (2000): 21.

24. Clarey, "Gender Test after Gold Medal Finish."

25. Jasbir Puar, *Terrorist Assemblages: Homonationalism in Queer Times* (Durham: Duke University Press, 2007), 183.

26. Pamela Scully, "The Trials of Caster Semenya," *The Defenders Online*, July 16, 2010, http://www.thedefendersonline.com/2010/07/16/the-trials-of-caster-semenya/.

27. John Berger, *Ways of Seeing* (New York: Penguin Books, 1990).

28. See Donna Haraway, "Situated Knowledges: The Science Question in Feminism and the Privilege of Partial Perspective," in *Simians, Cyborgs, and Women: The Reinvention of Nature* (New York: Routledge, 1991), 183–202.

29. Chandra Talpade Mohanty, "Under Western Eyes: Feminist Scholarship and Colonial Discourses," in *Third World Women and the Politics of Feminism*, ed. Chandra Talpade Mohanty, Ann Russo, and Lourdes Torres (Bloomington: Indiana University Press, 1991), 51–80.

30. See Fausto-Sterling, *Sexing the Body*.

31. Stanley I. Thangaraj, *Desi Hoop Dreams: Pickup Basketball and the Making of Asian-American Masculinity* (New York: New York University Press, 2015).

32. Gina Kolata, "I.O.C. Panel Calls for Treatment in Sex Ambiguity Cases," *New York Times*, January 20, 2010, http://www.nytimes.com/2010/01/21/sports/olympics/21ioc.html?hp.

33. Alice Dreger, "Where's the Rulebook for Sex Verification Testing?" *New York Times*, August 21, 2009, http://www.nytimes.com/2009/08/22/sports/22runner.html.

34. I. A. Hughes et al., "Consensus Statement on Management of Intersex Disorders," *Archives of Disease in Childhood* 91 (2005): 554–63.

35. Simon Hart, "Caster Semenya Tests 'Show High Testosterone Levels,'" *Telegraph*, August 24, 2009, http://www.telegraph.co.uk/sport/othersports/athletics/6078171/World-Athletics-Caster-Semenya-tests-show-high-testosterone-levels.html.

36. Dreger, "Where's the Rulebook for Sex Verification Testing?"

37. Dave Zirin and Sherry Wolf, "Caster Semenya: The Idiocy of Sex Testing," *The Nation*, August 21, 2009, http://www.thenation.com/doc/20090831/zirin_wolf.

38. Magubane, "Spectacles and Scholarship," 762.

39. Anita Brady, "'Could This Women's World Champ Be a Man?': Caster Semenya and the Limits of Being Human," *AntePodium* (2011): 4.

40. Zine Magubane, "Spectacles and Scholarship," 765.

41. Cheune, cited in David Smith, "Caster Semenya Row: 'Who Are White People to Question the Makeup of an African Girl? It Is Racism,'" *Guardian*, August 22, 2009, http://www.theguardian.com/sport/2009/aug/23/caster-semenya-athletics-gender.

42. Robyn Dixon, "Caster Semenya, Runner Subjected to Gender Test, Gets Hero's Welcome in South Africa," *Los Angeles Times*, August 26, 2009, http://articles.latimes.com/2009/aug/26/world/fg-africa-runner26.

43. Ibid.

44. Jacob Zuma, "IAAF Publicly Humiliated Caster Semenya," *Politics Web*, August 25, 2009, http://www.politicsweb.co.za/politicsweb/view/politicsweb/en/page71639?oid5140761&sn5Detail.

45. Malema, cited in Neville Hoad, "'Run, Caster Semenya, Run!': Nativism and the Translations of Gender Variance," *Safundi* 11, no. 4 (2010): 400.

46. Hoad, "'Run, Caster Semenya, Run!,'" 402.

47. Ibid.

48. Tavia Nyong'o, "The Unforgivable Transgression of Being Caster Semenya," *Women and Performance* 20, no. 1 (2010): 97.

49. Ibid., 96.

50. Brenna Munro, "Caster Semenya: Gods and Monsters," *Safundi* 11, no. 4 (2010): 384.

51. Magubane, "Spectacles and Scholarship," 771.

52. Ibid., 781.

53. Ibid., 782.

54. Ibid.

55. Ibid., 780.

56. Ibid., 768–69.

57. Lynnell Stephani Long, "Race and the Intersexed," *Windy City Times*, June 1, 2003, accessed January 25, 2016, http://www.windycitymediagroup.com/gay/lesbian/news/ARTICLE.php?AID=924; Dyana Bagby, "Atlanta Police Officer Shares Story of Being Intersex," *Southern Voice*, July 6, 2008, accessed January 25, 2016, http://transgriot.blogspot.com/2008/07/atlanta-police-officer-shares-story-of.html; *Intersexion*, directed by Grant Lahood (Wellington, New Zealand: Ponsby Productions Limited, 2011); and *One in 2000*, directed by Ajae Clearway (Champaign, IL: Polyvinyl Pictures, 2006).

58. Cynthia Krauss, "Hypospadias Surgery in a West African Context: The Surgical (Re-)construction of What?," *Feminist Theory* 14, no. 1 (2013): 83–103.

59. Mauro Cabral, "The Marks on Our Bodies," *Intersex Day*, October 25, 2015, http://intersexday.org/en/mauro-cabral-marks-bodies/.

60. Hilary Malatino, "Gone, Missing: Queering and Racializing Absence in Trans & Intersex Archives," in *Queer Feminist Science Studies: A Reader*, ed. Cyd Cipolla, Kristina Gupta, David A. Rubin, and Angela Willey (Seattle: University of Washington Press, forthcoming).

61. Fausto-Sterling, *Sexing the Body*.

62. Shari L. Dworkin, Amanda Lock Swarr, and Cheryl Cooky, "(In)Justice in Sport: The Treatment of South African Track Star Caster Semenya," *Feminist Studies*, 39, no. 1 (2013): 60.

63. Sally Gross, "Life in the Shadow of Gender," *The Witness*, August 29, 2009, http://www.news24.com/Archives/Witness/Life-in-the-shadow-of-gender-20150430.

64. Amanda Lock Swarr, *Sex in Transition: Remaking Gender and Race in South Africa* (Albany: State University of New York Press, 2012).

65. Caster Semenya, "Caster Semenya's Comeback Statement in Full," *The Guardian*, March 30, 2010, https://www.theguardian.com/sport/2010/mar/30/caster-semenya-comeback-statement.

66. For a brilliant analysis of the Chand controversy, see Ella Browning, "Rupturing the World of Elite Athletics: A Feminist Critical Discourse Analysis of the Suspension of the 2011 IAAF Regulations on Hyperandrogenism" (PhD diss., University of South Florida, 2016), http://scholarcommons.usf.edu/cgi/viewcontent.cgi?article=7385&context=etd.

67. Katrina Karkazis, "One Track Minds: Semenya, Chand, and the Violence of Public Scrutiny," *The Medium*, July 19, 2016, https://medium.com/@Karkazis/medias-one-track-mind-semenya-chand-the-violence-of-public-scrutiny-1aa6d1a08454#.kd88hug2t.

68. Chela Sandoval, *Methodology of the Oppressed* (Minneapolis: University of Minnesota Press, 2000).

69. Spivak, "Can the Subaltern Speak?"

70. Ben Smith, "Caster Semenya: 'What I Dream of Is to Become Olympic Champion,'" *BBC Sport/Athletics*, May 19, 2015, http://www.bbc.com/sport/0/athletics/32805695.

Conclusion

1. Jacques Derrida, "Choreographies," *Diacritics* 12, no. 2 (1982): 76.

2. I. A. Hughes et al., "Consensus Statement on Management of Intersex Disorders," *Archives of Disease in Childhood* 91 (2005): 554–63.

3. See DSD Consortium, "Clinical Guidelines for the Management of Disorders of Sex Development in Childhood," 2006, accessed January 29, 2010, http://www.dsdguidelines.org/htdocs/clinical/index.html; and DSD Consortium, "Handbook for Parents," 2006, accessed January 29, 2010, http://www.dsdguidelines.org/files/parents.pdf.

4. See the Accord Alliance website, accessed January 29, 2010, http://www.accordalliance.org/.

5. Elizabeth Reis, "Divergence or Disorder? The Politics of Naming Intersex," *Perspectives in Biology and Medicine* 50, no. 4 (2007): 535–43.

6. Ellen K. Feder and Katrina Karkazis, "What's in a Name? The Controversy over Disorders of Sex Development," *Hastings Center Report* 38, no. 5 (2008): 35.

7. Ellen K. Feder, "Imperatives of Normality: From 'Intersex' to 'Disorders of Sex Development,'" *GLQ* 15, no. 2 (2009): 227. For an extended elaboration of this argument, see Feder, *Making Sense of Intersex: Changing Ethical Perspectives in Biomedicine* (Bloomington: Indiana University Press, 2014).

8. Feder, *Making Sense of Intersex*, 5.

9. Feder, "Imperatives of Normality," 227.

10. In recent years, "human flourishing" has become a keyword in neoliberal discourses of sustainable development. These discourses help to secure the upward redistribution of wealth and the reproduction of vast inequalities between the global north and the global south. What structures of power define what gets to

count as human flourishing? Which groups' interests are served by this definition? See Lisa Duggan, *The Twilight of Equality? Neoliberalism, Cultural Politics, and the Attack on Democracy* (Boston: Beacon Press, 2004); David Harvey, *A Brief History of Neoliberalism* (Oxford: Oxford University Press, 2005); and Aihwa Ong, *Neoliberalism as Exception: Mutations in Citizenship and Sovereignty* (Durham: Duke University Press, 2006).

11. Feder, "Imperatives of Normality," 225.

12. Georgiann Davis, *Contesting Intersex: The Dubious Diagnosis* (New York: New York University Press, 2015), 21.

13. Morgan Holmes, "Mind the Gaps: Intersex and (Re-Productive) Spaces in Disability Studies and Bioethics," *Bioethical Inquiry* 5, nos. 2–3 (2008): 169–81; and Jennifer Germon, *Gender: A Genealogy of an Idea* (New York: Palgrave Macmillan, 2009).

14. Reis, "Divergence or Disorder," 538.

15. Elizabeth Reis, *Bodies in Doubt: An American History of Intersex* (Baltimore: Johns Hopkins University Press, 2012).

16. Reis, "Divergence or Disorder," 539.

17. See Cheryl Chase, "Hermaphrodites with Attitude: Mapping the Emergence of Intersex Political Activism," *GLQ* 4, no. 2 (1998): 189–211.

18. Jayne Caudwell, "Sex Watch: Surveying Women's Sexed and Gendered Bodies at the Olympics," in *Watching the Olympics: Politics, Power, and Representation*, ed. John Peter Sugden and Alan Tomlinson (New York: Routledge, 2012), 157.

19. See Susan B. Frampton and Patrick Charmel, eds., *Putting Patients First: Best Practices in Patient-Centered Care* (San Francisco: Jossey-Bass, 2009).

20. Marcus de Maria Arana (principal author), "A Human Rights Investigation into the Medical 'Normalization' of Intersex People: A Report of a Public Hearing by the Human Rights Commission of the City and County of San Francisco," April 28, 2005, accessed March 9, 2014, http://www.isna.org/files/SFHRC_Intersex_Report.pdf; and European Network of Legal Experts, "Trans and Intersex People: Discrimination on the Grounds of Sex, Gender Identity and Gender Expression," Luxembourg: European Union, 2011, accessed December 1, 2014, http://www.coe.int/t/dg4/lgbt/Source/trans_and_intersex_people_EC_EN.pdf.

21. Emily Grabham, "Citizen Bodies, Intersex Citizenship," *Sexualities* 10, no. 1 (2007): 40.

22. Ibid., 38.

23. Ibid., 40.

24. Ibid., 45.

25. See Laura Briggs, *Reproducing Empire: Race, Sex, Science, and U.S. Imperialism in Puerto Rico* (Berkeley: University of California Press, 2002).

26. Grabham, "Citizen Bodies, Intersex Citizenship," 40.

27. Katrina Karkazis, *Fixing Sex: Intersex, Medical Authority, and Lived Experience* (Durham: Duke University Press, 2008), 242.

28. Audre Lorde, *Sister Outsider: Essays and Speeches* (New York: The Crossing Press, 1984), 116.

29. On the will-to-knowledge, see Michel Foucault, *The History of Sexuality, Vol. 1: An Introduction*, trans., Robert Hurley (New York: Vintage, 1990).

30. Lisa Downing, Iain Morland, and Nikki Sullivan, *Fuckology: Critical Essays on John Money's Diagnostic Concepts* (Chicago: University of Chicago Press, 2015).

31. Ibid., 197.

32. See Jose Esteban Munoz, *Disidentifications: Queers of Color and the Performance of Politics* (Minneapolis: University of Minnesota Press, 1999).

33. Michel Foucault, *Discipline and Punish: The Birth of the Prison*, trans. Alan Sheridan (New York: Penguin, 1977).

34. Reis, "Divergence or Disorder," 541.

35. Anne Tamar-Mattis and Milton Diamond, "Managing Variations in Sex Development," *Journal of Pediatric Endocrinology and Metabolism* 20, no. 4 (2007): 552–53.

36. Anne Fausto-Sterling, *Sex/Gender: Biology in a Social World* (New York: Routledge, 2012), 2.

37. See Michael Anker, *Ethics of Uncertainty: Aporetic Openings* (New York: Atropos Press, 2009).

38. Jacques Derrida, *Rogues: Two Essays on Reason*, trans. Pascale-Anne Brault and Michael Nass (Stanford: Stanford University Press, 2005).

39. Thomas Keenan, *Fables of Responsibility: Aberrations and Predicaments in Ethics and Politics* (Stanford: Stanford University Press, 1997), 1–2.

40. Derrida, cited in J. Hillis Miller, *For Derrida* (New York: Fordham University Press, 2009), 26; original in Derrida, "Force of Law: The Mystical Foundation of Authority," trans. Mary Quaintance, in *Deconstruction and the Possibility of Justice*, ed. Drucilla Cornell, Michael Rosenfeld, and David Gray Carlson (New York: Routledge, 1992), 3–67.

41. Keenan, *Fables of Responsibility*, 7.

42. Gayatri Chakravorty Spivak, *A Critique of Postcolonial Reason: Toward a History of the Vanishing Present* (Cambridge: Harvard University Press, 1999).

43. See Inter/Act Youth, accessed December 15, 2015, http://interactyouth.org/; Thea Hillman, *Intersex (For Lack of a Better Word)* (San Francisco: Manic D Press, 2008); Del LaGrace Volcano, "The Herm Portfolio," *GLQ* 15, no. 2 (2009): 261–65; Ins A Kromminga, "1-0-1 [one 'o one] Intersex," 2005, accessed January 12, 2016, http://www.genderfreenation.de/art/index.html.

44. Lorde, *Sister Outsider*, 115.

45. Iain Morland, "Postmodern Intersex," in *Ethics and Intersex*, ed. Sharon E. Sytsma (Netherlands: Springer, 2006), 319–31.

46. On this concept, see Derrida, *Rogues*.

47. Fausto-Sterling, *Sex/Gender*, 123.

48. Donna Haraway, "The Cyborg Manifesto: Science, Technology, and Socialist-Feminism in the Late Twentieth Century," in *Simians, Cyborgs, and Women: The Reinvention of Nature* (New York: Routledge, 1991), 149–82.

Bibliography

18 U.S.C. § 116 (Supp. III 1997).

Abelove, Henry, Michele Aina Barale, and David M. Halperin, eds. *The Lesbian and Gay Studies Reader* (New York: Routledge, 1993).

Abu-Lughod, Lila. "Do Muslim Women Really Need Saving? Anthropological Reflections on Cultural Relativism and Its Others." *American Anthropologist* 104, no. 3 (2002): 783–90.

Abusharaf, Rogaia. "Unmasking Tradition." *Sciences* (March/April 1998): 23–27.

Accord Alliance. http://www.accordalliance.org/.

Advocates for Informed Choice. http://www.iiclaw.org/.

Agius, Silvan, Morgan Carpenter, and Dan Christian Ghattas. "Third Gender: A Step Toward Ending Intersex Discrimination." *Spiegel Online International.* August 22, 2013. http://www.spiegel.de/international/europe/third-gender-option-in-germany-a-small-step-for-intersex-recognition-a-917650.html.

Ahmed, Sara. "Imaginary Prohibitions: Some Preliminary Remarks on the Founding Gestures of the 'New Materialism.'" *European Journal of Women's Studies* 15, no. 1 (2008): 23–39.

Alaimo, Stacy. *Bodily Natures: Science, Environment, and the Material Self* (Bloomington: Indiana University Press, 2010).

Alarcón, Norma. "The Theoretical Subject(s) of *This Bridge called My Back* and Anglo-American Feminism." In *Haciendo Caras: Making Face, Making Soul: Creative and Critical Perspectives by Feminists of Color*, edited by Gloria Anzaldua, 356–69 (San Francisco: Aunt Lute Books, 1990).

Alcoff, Linda Martin. "The Problem of Speaking for Others." In *Who Can Speak? Authority and Critical Identity*, edited by Judith Roof and Robyn Wiegman, 97–119 (Urbana: University of Illinois Press, 1995).

Androgen Insensitivity Syndrome Support Group. http://www.aissg.org/INDEX.HTM.

Anker, Michael. *Ethics of Uncertainty: Aporetic Openings* (New York: Atropos Press, 2009).

Arana, Marcus, with the San Francisco Human Rights Commission. "A Human Rights Investigation into the Medical 'Normalization' of Intersex People: A Report of a Public Hearing by the Human Rights Commission of the City and County of San Francisco." 2005. http://www.sfgov.org/site/uploadedfiles/sfhumanrights/Committee_Meetings/Lesbian_ Gay_Bisexual_Transgender/SFHRC%20Intersex%20Report(1).pdf.

Atkins, Dawn, ed. *Looking Queer: Body Image and Gay Identity in Lesbian, Bisexual, Gay, and Transgender Communities* (New York: Harrington Park Press, 1998).

Australian Parliament. "Sex Discrimination Amendment (Sexual Orientation, Gender Identity and Intersex Status) Bill." http://www.aph.gov.au/Parliamentary_Business/Bills_Legislation/Bills_Search_Results/Result?bId=r5026.

Bagby, Dyana. "Atlanta Police Officer Shares Story of Being Intersex." *Southern Voice*. July 6, 2008. Accessed January 25, 2016. http://transgriot.blogspot.com/2008/07/atlanta-police-officer-shares-story-of.html.

Barad, Karen. *Meeting the Universe Halfway: Quantum Physics and the Entanglement of Matter and Meaning* (Durham: Duke University Press, 2007).

Barlow, Tani E. "'green blade in the act of being grazed': Late Capital, Flexible Bodies, Critical Intelligibility." *differences: A Journal of Feminist Cultural Studies* 10, no. 3 (1998): 119–58.

Basu, Amrita. "Globalization of the Local/Localization of the Global: Mapping Transnational Women's Movements." *Meridians: Feminism, Race, Transnationalism* 1, no. 1 (2000): 68–84.

Ben-Asher, Noah. "The Necessity of Sex Change: A Struggle for Intersex and Transsex Liberties." *Harvard Journal of Law and Gender* 29, no. 1 (2006): 51–98.

Bennett, Jane. *Vibrant Matter: A Political Ecology of Things* (Durham: Duke University Press, 2010).

Bennett-Smith, Meredith. "Mark and Pam Crawford, Parents of Intersex Child, Sue South Carolina for Sex Assignment Surgery." *The Huffington Post*, May 5, 2013. http://www.huffingtonpost.com/2013/05/15/mark-pam-crawford-intersex-child_n_3280353.html.

Berger, John. *Ways of Seeing* (New York: Penguin Books, 1990).

Berlant, Lauren. *The Queen of America Goes to Washington City: Essays on Sex and Citizenship* (Durham: Duke University Press, 1997).

Bettcher, Talia Mae. "Intersexuality, Transgender, and Transsexuality." In *Oxford Handbook of Feminist Theory*, edited by Mary Hawkesworth and Lisa Jane Disch, 407–27 (New York: Oxford University Press, 2016).

Blackless, Melanie, Anthony Charuvastra, Amanda Derryck, Anne Fausto-Sterling, Karl Lauzanne, and Ellen Lee. "How Sexually Dimorphic Are We? Review and Synthesis." *American Journal of Human Biology* 12, no. 2 (March/April 2000): 151–66.

Bloom, Amy. *Normal: Transsexual CEOs, Crossdressing Cops, and Hermaphrodites with Attitude* (New York: Vintage, 2003).

Bordo, Susan. *Unbearable Weight: Feminism, Western Culture, and the Body* (Berkeley: University of California Press, 2004).

Bornstein, Kate. *Gender Outlaw: On Men, Women, and the Rest of Us* (New York: Vintage, 1995).

Boxer, Marilyn Jacoby. *When Women Ask the Questions: Creating Women's Studies in America* (Baltimore: Johns Hopkins University Press, 1998).

Bradotti, Rosi. *The Posthuman* (Cambridge: Polity, 2013).

Brady, Anita. "'Could This Women's World Champ Be a Man?': Caster Semenya and the Limits of Being Human." *AntePodium* (2011): 1–16.

Breu, Christopher. *Insistence of the Material: Literature in the Age of Biopolitics* (Minneapolis: University of Minnesota Press, 2014).

Briggs, Laura. *Reproducing Empire: Race, Sex, Science, and U.S. Imperialism in Puerto Rico* (Berkeley: University of California Press, 2002).

Briggs, Laura, Gladys McCormick, and J. T. Way. "Transnationalism: A Category of Analysis." *American Quarterly* 60, no. 3 (September 2008): 625–48.

Brown, Wendy. "Suffering the Paradoxes of Rights." In *Left Legalism/Left Critique*, edited by Wendy Brown and Janet Halley, 420–34 (Durham: Duke University Press, 2002).

Brown, Wendy, and Janet Halley, eds. *Left Legalism/Left Critique* (Durham: Duke University Press, 2002), 420–34.

Browning, Ella. "Rupturing the World of Elite Athletics: A Feminist Critical Discourse Analysis of the Suspension of the 2011 IAAF Regulations on Hyperandrogenism." PhD diss., University of South Florida, 2016. http://scholarcommons.usf.edu/cgi/viewcontent.cgi?article=7385&context=etd.

Burchell, Graham, Colin Gordon, and Peter Miller, eds. *The Foucault Effect: Studies in Governmentality* (Chicago: University of Chicago Press, 1991).

Butler, Judith. *Gender Trouble: Feminism and the Subversion of Identity* (New York: Routledge, 1990). Reprinted as *Gender Trouble: Feminism and the Subversion of Identity*, 10th Anniversary Edition (New York: Routledge, 1999).

———. *Bodies That Matter: On the Discursive Limits of "Sex"* (New York: Routledge, 1993).

———. "Critically Queer." *GLQ* 1, no. 1 (1993): 17–32.

———. "Against Proper Objects." *differences* 6, nos. 2/3 (1994): 1–26.

———. "Doing Justice to Someone: Sex Reassignment and Allegories of Transsexuality." *GLQ* 7, no. 4 (2001): 621–36. Reprinted in Butler, *Undoing Gender* (New York: Routledge, 2002), 57–76.

———. *Undoing Gender* (New York: Routledge, 2004).

Butler, Judith, John Guillory, and Kendall Thomas, eds. *What's Left of Theory?* (New York: Routledge, 2000).

Cabral, Mauro. "The Marks on Our Bodies." *Intersex Day* (October 25, 2015). http://intersexday.org/en/mauro-cabral-marks-bodies/.

Califia, Pat. *Sex Changes: The Politics of Transgenderism* (San Francisco: Cleis Press, 1997).

Canguilhem, Georges. *The Normal and the Pathological*. Translated by Carolyn R. Fawcett (London: Zone Books, 1991).

Castellanos, Angela. "Colombia Confronts Female Genital Mutilation." *HR Reality Check*. August 18, 2008. Accessed July 14, 2014. http://rhrealitycheck.org/article/2008/08/18/colombia-confronts-female-genital-mutilation/.

Caudwell, Jayne. "Sex Watch: Surveying Women's Sexed and Gendered Bodies at the Olympics." In *Watching the Olympics: Politics, Power, and Representation*, edited by John Peter Sugden and Alan Tomlinson, 151–64 (New York: Routledge, 2012).

Chakrabarty, Dipesh. *Provincializing Europe: Postcolonial Thought and Historical Difference* (Princeton: Princeton University Press, 2007).

Chase, Cheryl. "Intersexual Rights," *Sciences* (July/August 1993): 3.

———. "Hermaphrodites with Attitude: Mapping the Emergence of Intersex Political Activism." *GLQ* 4, no. 2 (1998): 189–211.

———. "Affronting Reason." In *Looking Queer: Body Image and Gay Identity in Lesbian, Bisexual, Gay, and Transgender Communities*, edited by Dawn Atkins, 205–19 (New York: Harrington Park Press, 1998). Reprinted in Joan Nestle, Clare Howell, and Riki Wilchineds, *GenderQueer: Voices from Beyond the Sexual Binary* (Los Angeles: Alyson Books, 2002), 204–19.

———. " 'Cultural Practice' or 'Reconstructive Surgery'? U.S. Genital Cutting, the Intersex Movement, and Medical Double Standards." In *Genital Cutting and Transnational Sisterhood: Disputing U.S. Polemics*, edited by Stanlie M. James and Clair C. Robertson, 144–45 (Urbana: University of Illinois Press, 2002).

———. "What Is the Agenda of the Intersex Patient Advocacy Movement?" *Endocrinologist* 13 (2003): 240–42.

Chen, Mel. *Animacies: Biopolitics, Racial Mattering, and Queer Affect* (Durham: Duke University Press, 2012).

Cheah, Pheng. "Mattering." *Diacritics* 26, no. 1 (1996): 108–39.

———. *Inhuman Conditions: On Cosmopolitanism and Human Rights* (Cambridge: Harvard University Press, 2007).

Cipolla, Cyd, Kristina Gupta, David A. Rubin, and Angela Willey, eds. *Queer Feminist Science Studies: A Reader* (Seattle: University of Washington Press, forthcoming).

Clarey, Christopher. "Gender Test after Gold Medal Finish." *New York Times*, August 19, 2009. http://www.nytimes.com/2009/08/20/sports/20runner.html.

Colapinto, John. *As Nature Made Him: The Boy Who Was Raised a Girl* (New York: HarperCollins, 2000).

Colligan, Sumi. "Why the Intersexed Shouldn't Be Fixed: Insights from Queer Theory and Disability Studies." In *Gendering Disability*, edited by Bonnie G. Smith and Beth Hutchison, 45–60 (New Brunswick: Rutgers University Press, 2004).

Collins, Patricia Hill. *Black Feminist Thought: Knowledge, Consciousness, and the Politics of Empowerment* (New York: Routledge, 1990).

———. *Fighting Words: Black Women and the Search for Justice* (Minneapolis: University of Minnesota Press, 1998).

———. "Defining Black Feminist Thought." In *Feminist Theory Reader: Local and Global Perspectives*, edited by Carole R. McCann and Seung-kyung Kim, 379–94 (New York: Routledge, 2013).

Combahee River Collective. "A Black Feminist Statement." In *Feminist Theory Reader: Local and Global Perspectives*, edited by Carole R. McCann and Seung-kyung Kim, 116–22 (New York: Routledge, 2013).

Coole, Dianna, and Samantha Frost, eds. *New Materialisms: Ontology, Agency, and Politics* (Durham: Duke University Press, 2010).

Constitutional Court of Colombia. "The Rights of Intersexed Infants and Children: Decision of the Colombian Constitutional Court, Bogota, Colombia, 12 May 1999 (SU-337/99)." In *Transgender Rights*, edited by Paisley Currah, Richard M. Juang, and Shannon Price Minter and translated by Nohemy Solozano-Thompson, 122–40 (Minneapolis: University of Minnesota Press, 2006).

———. Sentencia No. T-477/95 (1999), 5.2.1.

Coole, Diana, and Samantha Frost, eds. *New Materialisms: Ontology, Agency, and Politics* (Durham: Duke University Press, 2010).

Coventry, Martha. "Making the Cut." *Ms. Magazine*, October–November 2000. http://www.msmagazine.com/oct00/makingthecut.html.

Creighton, Sarah M., Catherine L. Minto, and Stuart J. Steele. "Objective Cosmetic and Anatomical Outcomes at Adolescence of Feminising Surgery for Ambiguous Genitalia Done in Childhood." *Lancet* 358 (2001): 124–25.

Crenshaw, Kimberle. "Demarginalizing the Intersection of Race and Sex: A Black Feminist Critique of Antidiscrimination Doctrine, Feminist Theory and Antiracist Politics." *University of Chicago Legal Forum* 140 (1989): 139–67.

———. "Mapping the Margins: Intersectionality, Identity Politics, and Violence against Women of Color." *Stanford Law Review* 43 (1991): 1241–52.

Crespo, Julia Sandra Bernal. "Estados Intersexuales en Menores de Edad: Los Principios de Autonomía y Beneficencia." ["Intersex States in Minors: The Principles of Autonomy and Beneficence."] *Revista de Derecho* 36 (2011): 53–86.

Crouch, N. S., C. L. Minto, L.-M. Laio, C. R. J. Woodhouse, and S. M. Creighton. "Genital Sensation after Feminizing Genitoplasty for Congenital Adrenal Hyperplasia: A Pilot Study." *BJU International* 93 (2004): 135–38.

Cruikshank, Barbara. *The Will to Empower: Democratic Citizens and Other Subjects* (Ithaca: Cornell University Press, 1999).

Currah, Paisley, Richard M. Juang, and Shannon Price Minter. "Introduction." In *Transgender Rights*, edited by Paisley Currah, Richard M. Juang, and Shannon Price Minter, xii–xxiv (Minneapolis: University of Minnesota Press, 2006).

Cvetkovich, Ann. *An Archive of Feelings: Trauma, Sexuality, and Lesbian Public Cultures* (Durham: Duke University Press, 2003).

Daly, Mary. *Gyn/Ecology: The Metaethics of Radical Feminism* (Boston: Beacon Press, 1978).

Daston, Lorraine, and Katherine Park. "The Hermaphrodite and the Orders of Nature: Sexual Ambiguity in Early Modern France." *GLQ* 1, no. 4 (1995): 419–38.

Davis, Georgiann. *Contesting Intersex: The Dubious Diagnosis* (New York: New York University Press, 2015).

Davis, Kathy. *The Making of* Our Bodies, Ourselves: *How Feminism Travels Across Borders* (Durham: Duke University Press, 2007).

Davis, Lennard. *Enforcing Normalcy: Disability, Deafness, and the Body* (London: Verso, 1995).

De Lauretis, Teresa. *Technologies of Gender: Essays on Theory, Film, and Fiction* (Bloomington: Indiana University Press, 1987).

Derrida, Jacques. *Of Grammatology*. Translated by Gayatri Chakravorty Spivak (Baltimore: Johns Hopkins University Press, 1974).

———. "Choreographies." *Diacritics* 12, no. 2 (1982): 66–76.

———. *Limited Inc.* Translated by Jeffrey Mehlmen and Samuel Weber (Evanston: Northwestern University Press, 1988).

———, translated by Mary Quaintance. "Force of Law: The Mystical Foundation of Authority." In *Deconstruction and the Possibility of Justice*, edited by Drucilla Cornell, Michael Rosenfeld, and David Gray Carlson, 3–67 (New York: Routledge, 1992).

———. *Specters of Marx: The State of the Debt, the Work of Mourning and the New International*. Translated by Peggy Kamuf (New York: Routledge, 1994).

———. *Monolingualism of the Other; or, The Prosthesis of Origin*. Translated by Patrick Mensah (Stanford: Stanford University Press, 1998).

———. *Rogues: Two Essays on Reason*. Translated by Pascale-Anne Brault and Michael Nass (Stanford: Stanford University Press, 2005).

Deutscher, Penelope. *The Philosophy of Simone de Beauvoir: Ambiguity, Conversion, Resistance* (Cambridge: Cambridge University Press, 2008).

Diamond, M., and H. K. Sigmundson. "Sex Reassignment at Birth: Long-term Review and Implications." *Archives of Pediatric and Adolescent Medicine* 151 (October 1997): 298–304.

Dixon, Robyn. "Caster Semenya, Runner Subjected to Gender Test, Gets Hero's Welcome in South Africa." *Los Angeles Times*, August 26, 2009. http://articles.latimes.com/2009/aug/26/world/fg-africa-runner26.

Doell, Ruth G., and Helen E. Longino. "Sex Hormones and Human Behavior: A Critique of the Linear Model." *Journal of Homosexuality* 15 (1988): 55–78.

Downing, Lisa, Iain Morland, and Nikki Sullivan. *Fuckology: Critical Essays on John Money's Diagnostic Concepts* (Chicago: University of Chicago Press, 2015).

Doyle, Jennifer. "Dirt Off Her Shoulders." *GLQ* 19, no. 4 (2013): 419–33.

Dreger, Alice, ed. *Intersex in the Age of Ethics* (Hagerstown: University Publishing Group, 1999).

———. *Hermaphrodites and the Medical Invention of Sex* (Cambridge: Harvard University Press, 2000).

———. "Intersex and Human Rights: The Long View." In *Ethics and Intersex*, edited by Sharon E. Sytsma, 73–86 (Netherlands: Springer, 2006).

———. "Why 'Disorders of Sex Development'? (On Language and Life)." 2007. http://www.alicedreger.com/dsd.html.

———. "Where's the Rulebook for Sex Verification Testing?" *New York Times*, August 21, 2009. http://www.nytimes.com/2009/08/22/sports/22runner.html

———. "Shifting the Paradigm of Intersex Treatment." http://www.isna.org/compare.

———. "Ending Forced 'Genital-Normalizing' Surgeries." *The Atlantic*, February 25, 2013. http://www.theatlantic.com/health/archive/2013/02/ending-forced-genital-normalizing-surgeries/273300/.

Dreger, Alice D., and April M. Herndon, "Progress and Politics in the Intersex Rights Movement: Feminist Theory in Action." *GLQ* 15, no. 2 (2009): 199–224.

DSD Consortium. "Clinical Guidelines for the Management of Disorders of Sex Development in Childhood." 2006. http://www.dsdguidelines.org/htdocs/clinical/index.html.

———. "Handbook for Parents." 2006. http://www.dsdguidelines.org/files/parents.pdf.

Duarte, Ángela Ixkic Bastian. "From the Margins of Latin American Feminism: Indigenous and Lesbian Feminisms." *Signs: Journal of Women in Culture and Society* 38, no. 1 (2012): 153–78.

Duggan, Lisa. *The Twilight of Equality? Neoliberalism, Cultural Politics, and the Attack on Democracy* (Boston: Beacon Press, 2003).

Dworkin, Shari L., Amanda Lock Swarr, and Cheryl Cooky. "(In)Justice in Sport: The Treatment of South African Track Star Caster Semenya." *Feminist Studies* 39, no. 1 (2013): 40–69.

Edelman, Lee. *No Future: Queer Theory and the Death Drive* (Durham: Duke University Press, 2004).

Ehrenreich, Barbara, and Deirdre English. *For Her Own Good: Two Centuries of the Experts' Advice to Women* (New York: Anchor Books, 2005).

Elam, Diane. *Feminism and Deconstruction: Ms. en Abyme* (New York: Routledge, 1994).

Ellis, Havelock. *Sexual Inversion* (London: University Press, 1897).

———. *Studies in the Psychology of Sex*, 2 vols. (New York: Random House, 1905).

European Network of Legal Experts. "Trans and Intersex People: Discrimination on the Grounds of Sex, Gender Identity and Gender Expression." Luxembourg: European Union, 2011. http://www.coe.int/t/dg4/lgbt/Source/trans_and_intersex_people_EC_EN.pdf.

Fausto-Sterling, Anne. *Myths of Gender: Biological Theories about Women and Men* (New York: Basic Books, 1985).

———. "The Five Sexes: Why Male and Female Are Not Enough." *Sciences* (March–April 1993): 20–24.

———. "The Five Sexes, Revisited." *Sciences* 40, no. 4 (2000): 19–23.

———. *Sexing the Body: Gender Politics and the Construction of Sexuality* (New York: Basic Books, 2000).

———. *Sex/Gender: Biology in a Social World* (New York: Routledge, 2012).

Feder, Ellen K. "Imperatives of Normality: From 'Intersex' to 'Disorders of Sex Development.'" *GLQ* 15, no. 2 (2009): 225–47.

———. *Making Sense of Intersex: Changing Ethical Perspectives in Biomedicine* (Bloomington: Indiana University Press, 2014).

Feder, Ellen K., and Katrina Karkazis, "What's in a Name? The Controversy over Disorders of Sex Development." *Hastings Center Report* 38, no. 5 (2008): 33–35.

Feminist Majority Foundation. "Medical Violence against Intersex Individuals in the United States." http://www.feministcampus.org/fmla/printable-materials/v-day05/intersex_activism.pdf.

Feray, Jean-Claude, and Manfred Herzer. "Homosexual Studies and Politics in the 19th Century: Karl Maria Kertbeny." *Journal of Homosexuality* 19, no. 1 (1990): 23–48.

Ferguson, Roderick. *Aberrations in Black: Toward a Queer of Color Critique* (Minneapolis: University of Minnesota Press, 2003).

———. *The Reorder of Things: The University and Its Pedagogies of Minority Difference* (Minneapolis: University of Minnesota Press, 2012).

Fernandes, Leela. *Transnational Feminism in the United States: Knowledge, Ethics, Power* (New York: New York University Press, 2013).

Foucault, Michel. *Discipline and Punish: The Birth of the Prison*. Translated by Alan Sheridan (New York: Penguin, 1977).

———. *Herculine Barbin dite Alexina B.* (Paris: Gallimard, 1978).

———. *Herculine Barbin: Being the Recently Discovered Memoirs of a Nineteenth-Century French Hermaphrodite* (New York: Pantheon, 1980).

———. "Nietzsche, Genealogy, History." In *The Foucault Reader*, edited by Paul Rabinow, 76–100 (New York: Pantheon Books, 1984).

———. *The Order of Things: An Archeology of the Human Sciences* (New York: Routledge, 1989).

———. *The History of Sexuality, Vol. 1: An Introduction* (New York: Vintage, 1990).

———. *The History of Sexuality, Volume II: The Use of Pleasure*. Translated by Robert Hurley (New York: Vintage Books, 1990).

———. *Dits et Écrits: 1976–1988* (Paris: Gallimard, 2001), 934–42.

———. "Le Vrai sexe." *Arcadie* 323 (Novembre 1980): 617–25. Reprinted in Foucault, *Dits et Écrits: 1976–1988* (Paris: Gallimard, 2001), 934–42.

———. *Abnormal: Lectures at the Collège de France, 1974–1975*. Translated by Graham Burchell (New York: Picador, 2004).

———. "22 January 1975." In *Abnormal: Lectures at the College de France*, translated by Graham Burchell 1974–1975, 55–81 (New York: Picador, 2004).

———. *History of Madness* (New York: Routledge, 2006).

Frampton, Susan B., and Patrick Charmel, eds. *Putting Patients First: Best Practices in Patient-Centered Care* (San Francisco: Jossey-Bass, 2009).

Freeman, Carla. *High Tech and High Heels in the Global Economy: Women, Work, and Pink-Collar Identities in the Caribbean* (Durham: Duke University Press, 2000).

Fuss, Diana. *Essentially Speaking: Feminism, Nature, and Difference* (New York: Routledge, 1989).

Garland-Thomson, Rosemarie. *Extraordinary Bodies: Figuring Physical Disability in American Culture and Literature* (New York: Columbia University Press, 1997).

Germon, Jennifer. *Gender: A Genealogy of an Idea* (New York: Palgrave Macmillan, 2009).

Ghattas, Dan Christian. *Human Rights Between the Sexes: A Preliminary Study on the Life Situations of Inter* Individuals*, (Berlin: Heinrich-Boll-Stiftung, 2013).

Gilman, Sander L. *Difference and Pathology: Stereotypes of Sexuality, Race, and Madness* (Ithaca: Cornell University Press, 1985).

Gordon, Avery. *Ghostly Matters: Haunting and the Sociological Imaginary* (Minneapolis: University of Minnesota Press, 2008).

Gould, Stephen Jay. *The Mismeasure of Man* (New York: W.W. Norton, 1996).

Grabham, Emily. "Citizen Bodies, Intersex Citizenship." *Sexualities* 10, no. 1 (2007): 29–48.

Grandin, Greg. *Empire's Workshop: Latin America, The United States, and the Rise of the New Imperialism* (New York: Holt Paperbacks, 2006).

Greenberg, Julie A. *Intersexuality and the Law: Why Sex Matters* (New York: New York University Press, 2012).

Greenberg, Julie A., and Cheryl Chase. "Background of the Colombia Decisions." 1999. http://www.isna.org/node/21.

Grewal, Inderpal. *Transnational America: Feminisms, Diasporas, Neoliberalisms* (Durham: Duke University Press, 2005).

Grewal, Inderpal, and Caren Kaplan, eds. *Scattered Hegemonies: Postmodernity and Transnational Feminist Practices* (Minneapolis: University of Minnesota Press, 1994).

———. *An Introduction to Women's Studies: Gender in a Transnational World.* 2nd edition (New York: McGraw-Hill, 2002).

———. "Transnational Practices and Interdisciplinary Feminist Scholarship: Refiguring Women's and Gender Studies." In *Women's Studies on Its Own*, edited by Robyn Wiegman, 66–81 (Durham: Duke University Press, 2002).

Gross, Sally. "Life in the Shadow of Gender." *The Witness*. August 29, 2009. http://www.news24.com/Archives/Witness/Life-in-the-shadow-of-gender-20150430.

Grusin, Richard, ed. *The Nonhuman Turn* (Minneapolis: University of Minnesota Press, 2015).

Halberstam, Judith. *Female Masculinity* (Durham: Duke University Press, 1998).

———. *In a Queer Time and Place: Transgender Bodies, Subcultural Lives* (Durham: Duke University Press, 2005).

Halley, Janet. "'Like Race' Arguments." In *What's Left of Theory?*, edited by Judith Butler, John Guillory, and Kendall Thomas, 40–74 (New York: Routledge, 2000).

Haraway, Donna J. "The Cyborg Manifesto: Science, Technology, and Socialist-Feminism in the Late Twentieth Century." In *Simians, Cyborgs, and Women: The Reinvention of Nature*, 149–82 (New York: Routledge, 1991).

———. *Simians, Cyborgs, and Women: The Reinvention of Nature* (New York: Routledge, 1991).

———. "Situated Knowledges: The Science Question in Feminism and the Privilege of Partial Perspective." In *Simians, Cyborgs, and Women: The Reinvention of Nature* (New York: Routledge, 1991), 183–202.

———. *When Species Meet* (Minneapolis: University of Minnesota Press, 2007).

Harding, Sandra. *The Science Question in Feminism* (Ithaca: Cornell University Press, 1986).

Hart, Simon. "Caster Semenya Tests 'Show High Testosterone Levels.'" *Telegraph*, August 24, 2009. http://www.telegraph.co.uk/sport/othersports/athletics/6078171/World-Athletics-Caster-Semenya-tests-show-high-testosterone-levels.html.

Hartsock, Nancy C. M. "The Feminist Standpoint: Developing the Ground for a Specifically Feminist Historical Materialism." In *Discovering Reality: Feminist Perspectives on Epistemology, Metaphysics, Methodology, and Philosophy of Science*, edited by Sandra Harding and Merrill B. Hintikka, 283–310 (Dordrecht, Holland: Springer, 1983).

Harvey, David. *A Brief History of Neoliberalism* (Oxford: Oxford University Press, 2005).

Hausman, Bernice. *Changing Sex: Transsexualism, Technology, and the Idea of Gender* (Durham, Duke University Press, 1995).

Hegarty, Peter. *Gentlemen's Disagreement: Alfred Kinsey, Lewis Terman, and the Sexual Politics of Smart Men* (Chicago: University of Chicago Press, 2013).

Hemmings, Claire. *Why Stories Matter: The Political Grammar of Feminist Theory* (Durham: Duke University Press, 2011), 224.

Herndon, April. "Why Doesn't ISNA Want to Eradicate Gender?" 2006. http://www.isna.org/faq/not_eradicating_gender.

Herdt, Gilbert, ed. *Third Sex, Third Gender: Beyond Sexual Dimorphism in History and Culture* (London: Zone Books, 1996).

Hird, Myra. "Gender's Nature: Intersexuality, Transsexualism and the 'Sex'/'Gender' Binary." *Feminist Theory* 1, no. 3 (2000): 347–63.

Hirsch, Marianne, and Evelyn Fox Keller, eds. *Conflicts in Feminism* (New York: Routledge, 1990).

Hoad, Neville. " 'Run, Caster Semenya, Run!': Nativism and the Translations of Gender Variance." *Safundi* 11, no. 4 (2010): 397–405.

Holmes, Morgan. "Queer Cut Bodies." In *Queer Frontiers: Millennial Geographies, Genders and Generations*, edited by Joseph A Boone et. al., 84–110 (Madison: University of Wisconsin Press, 2000).

———. "Deciding Fate or Protecting a Developing Autonomy? Intersex Children and the Colombian Constitutional Court." In *Transgender Rights*, edited by Paisley Currah, Richard M. Juang, and Shannon Price Minter, 102–21 (Minneapolis: University of Minnesota Press, 2006).

———. *Intersex: A Perilous Difference* (Selinsgrove: Susquehanna University Press, 2008).

———. "Mind the Gaps: Intersex and (Re-Productive) Spaces in Disability Studies and Bioethics." *Bioethical Inquiry* 5, nos. 2–3 (2008): 169–81.

———, ed. *Critical Intersex* (London: Ashgate, 2009).

Hood-Williams, John. "Goodbye to Sex and Gender." *Sociological Review* 44, no. 1 (1996): 1–16.

Hong, Grace Kyungwon. *The Ruptures of American Capital: Women of Color Feminism and the Culture of Immigrant Labor* (Minneapolis: University of Minnesota Press, 2006).

Hong, Grace Kyungwon, and Roderick Ferguson, eds. *Strange Affinities: The Gender and Sexual Politics of Comparative Racializiation* (Durham: Duke University Press, 2011).

hooks, bell. *Feminist Theory: From Margin to Center* (Boston: South End Press, 1984).

Huffer, Lynne. *Mad for Foucault: Rethinking the Foundations of Queer Theory* (New York: Columbia University Press, 2010).

Hughes, I. A. et al. "Consensus Statement on Management of Intersex Disorders." *Archives of Disease in Childhood* 91 (2005): 554–63.

Intersex Society of North America. http://isna.org.

———. "ISNA's Amicus Brief on Intersex Genital Surgery." 1998. http://www.isna.org/node/97.

———. "Colombia High Court Restricts Surgery on Intersex Children." ISNA Press Release. 1999. http://www.isna.org/colombia/.

———. "ISNA Honored with Human Rights Award." ISNA Press Release. 1999. http://www.isna.org/node/15.

———. "Intersex Declared a Human Rights Issue." ISNA Press Release. 2005. http://www.isna.org/node/840.

Intersexion. Directed by Grant Lahood. Wellington, New Zealand: Ponsby Productions Limited, 2011.

James, Stanlie M., and Claire C. Roberston, eds. *Genital Cutting and Transnational Sisterhood: Disputing U.S. Polemics* (Urbana: University of Illinois Press, 2002).

Johnson, Barbara. *A World of Difference* (Baltimore: Johns Hopkins University Press, 1987).

Jordan-Young, Rebecca. *Brain Storm: The Flaws in the Science of Sex Differences* (Cambridge: Harvard University Press, 2011).

Joseph, Miranda. *Against the Romance of Community* (Minneapolis: University of Minnesota Press, 2002).

———."Analogy and Complicity: Women's Studies, Lesbian/Gay/Bisexual Studies, and Capitalism." In *Women's Studies on Its Own*, edited by Robyn Wiegman, 267–92 (Durham: Duke University Press, 2002).

Joseph, Miranda, and David Rubin. "Promising Complicities: On the Sex, Race and Globalization Project." In *The Blackwell Companion to Lesbian, Gay, Bisexual, Transgender, and Queer Studies*, edited by George E. Haggerty and Molly McGarry, 430–51 (London: Blackwell, 2007).

Kaplan, Caren. *Questions of Travel: Postmodern Discourses of Displacement* (Durham: Duke University Press, 1996).

Karkazis, Katrina. *Fixing Sex: Intersex, Medical Authority, and Lived Experience* (Durham: Duke University Press, 2008).

———. "One Track Minds: Semenya, Chand, and the Violence of Public Scrutiny." *The Medium*. July 19, 2016. https://medium.com/@Karkazis/medias-one-track-mind-semenya-chand-the-violence-of-public-scrutiny-1aa6d1a08454#.kd88hug2t.

Keenan, Thomas. *Fables of Responsibility: Aberrations and Predicaments in Ethics and Politics* (Stanford: Stanford University Press, 1997).

Kerry, Stephen. "'Intersex Imperialism' and the Case of Caster Semenya: The Unacceptable Woman's Body." *Scan: Journal of Media, Arts, Culture* 8, no. 1 (2011). Accessed July 1, 2014. http://scan.net.au/scan/journal/display.php?journal_id=158.

Kessler, Suzanne J. "The Medical Construction of Gender: Case Management of Intersexed Infants." *Signs* 16, no. 1 (1990): 33–38.

———. *Lessons from the Intersexed* (New Brunswick: Rutgers University Press, 1998).

Kimmel, Michael. *Guyland: The Perilous World Where Boys Become Men* (New York: Harper, 2006).

Koyama, Emi, and Lisa Weasel. "From Social Construction to Social Justice: Transforming How We Teach about Intersexuality." *Women's Studies Quarterly* 30, nos. 3–4 (Fall 2002): 169–78.

Krafft-Ebing, Richard von. *Psychopathia Sexualis* (Memphis: General Books LLC, 2009).

Kobewka, Scott. "Embera Vow to Stop Female Genital Mutilation." *Colombia Reports*. November 23, 2010. Accessed July 15, 2014. http://colombiareports.co/embera-vow-stop-female-genital-mutilation/.

Kolata, Gina. "I.O.C. Panel Calls for Treatment in Sex Ambiguity Cases." *New York Times*, January 20, 2010. http://www.nytimes.com/2010/01/21/sports/olympics/21ioc.html?hp.

Krauss, Cynthia. "Hypospadias Surgery in a West African Context: The Surgical (Re-)Construction of What?" *Feminist Theory* 14, no. 1 (2013): 83–103.

Kroker, Arthur. *Body Drift: Butler, Hayles, Haraway* (Minneapolis: University of Minnesota Press, 2012).

Kromminga, Ins A. "1-0-1 [one 'o one] Intersex." 2005. http://www.genderfreenation.de/art/index.html.

Kulick, Don. *Travesti: Sex, Gender, and Culture among Brazilian Transgendered Prostitutes* (Chicago: University of Chicago Press, 1998).

Laqueur, W. Thomas. "Notes from the (Non)Field: Teaching and Theorizing Women of Color." In *Women's Studies on Its Own*, edited by Robyn Wiegman, 82–105 (Durham: Duke University Press, 2002).

Leong, Karen J., Roberta Chevrette, Ann Hibner Koblitz, Karen Kuo, and Heather Switzer. "Introduction." *Frontiers: A Journal of Women's Studies* 36, no. 3 (2015): vii–xv.

Lewis, Vek. *Crossing Sex and Gender in Latin America* (New York: Palgrave MacMillan, 2010).

Long, Lynnell Stephani. "Race and the Intersexed." *Windy City Times*. June 1, 2003. Accessed January 25, 2016. http://www.windycitymediagroup.com/gay/lesbian/news/ARTICLE.php?AID=924.

Lorber, Judith. *Paradoxes of Gender* (New Haven: Yale University Press, 1995).

Lorde, Audre. *Sister Outsider: Essays and Speeches* (New York: Crossing Press, 1984).

Lugones, María. "Heterosexualism and the Colonial/Modern Gender System." *Hypatia* 22, no. 1 (2007): 186–209.

Magubane, Zine. "Spectacles and Scholarship: Caster Semenya, Intersex Studies, and the Problem of Race in Feminist Theory." *Signs* 39, no. 3 (2014): 761–85.

Mahmood, Saba. *Politics of Piety: The Islamic Revival and the Feminist Subject* (Princeton: Princeton University Press, 2005).

Malatino, Hilary. "Situating Bio-Logic, Refiguring Sex: Intersexuality and Coloniality." In *Critical Intersex*, edited by Morgan Holmes, 73–96 (Farnham: Ashgate, 2009).

———. "Gone, Missing: Queering and Racializing Absence in Trans & Intersex Archives." In *Queer Feminist Science Studies: A Reader*, ed. Cyd Cipolla, Kristina Gupta, David A. Rubin, and Angela Willey (Seattle: University of Washington Press, forthcoming).

Markowitz, Sally. "Pelvic Politics: Sexual Dimorphism and Racial Difference." *Signs* 26, no. 2 (2001): 389–414.

Martin, Luther H., Huck Gutman, and Patrick H. Hutton, eds. *Technologies of the Self: A Seminar with Michel Foucault* (Amherst: University of Massachusetts Press, 1988).

Mattingly, Cheryl, and Linda C. Garro, eds. *Narrative and the Cultural Construction of Illness and Healing* (Berkeley: University of California Press, 2000).

May, Vivian. *Pursuing Intersectionality, Unsettling Dominant Imaginaries* (New York: Routledge, 2015).

Mbembe, Achille. *On the Postcolony* (Berkeley: University of California Press, 2001).

McCall, Leslie. "The Complexity of Intersectionality." *Signs* 30, no. 3 (2005): 1771–800.

McCann, Carole R., and Seung-Kyung Kim, eds. *Feminist Theory Reader: Local and Global Perspectives* (New York: Routledge, 1993).

McRuer, Robert. *Crip Theory: Cultural Signs of Queerness and Disability* (New York: New York University Press, 2006).

Messer-Davidow, Ellen. *Disciplining Feminism: From Social Activism to Academic Discourse* (Durham: Duke University Press, 2002).

Meyerowitz, Joanne. *How Sex Changed: A History of Transsexuality in the United States* (Cambridge: Harvard University Press, 2002).

Miller, J. Hillis. *For Derrida* (New York: Fordham University Press, 2009).

Minh-ha, Trinh T. *Woman, Native, Other: Writing Postcoloniality and Feminism* (Bloomington: Indiana University Press, 1989).

Mohanty, Chandra Talpade. "Under Western Eyes: Feminist Scholarship and Colonial Discourses." In *Third World Women and the Politics of Feminism*, edited by Chandra Talpade Mohanty, Ann Russo, and Lourdes Torres, 51–80 (Bloomington: Indiana University Press, 1991).

Money, John. "Hermaphroditism: An Inquiry into the Nature of a Human Paradox." PhD diss., Harvard University, 1952.

———. "Hermaphroditism, Gender and Precocity in Hyperadrenocorticism: Psychological Findings." *Bulletin of the Johns Hopkins Hospital* 96 (1955): 253–64.

———. *Love and Love Sickness: The Science of Sex, Gender Difference, and Pair Bonding* (Baltimore: Johns Hopkins University Press, 1980).

———. *Gendermaps: Social Constructionism, Feminism, and Sexosophical History* (New York: Continuum, 1995).

———. "Lexical History and Constructionist Ideology of Gender." In *Gendermaps: Social Constructionism, Feminism, and Sexosophical History* (New York: Continuum, 1995), 15–32.

Money, John, and Anke E. Ehrhardt. *Man & Woman, Boy & Girl: The Differentiation and Dimorphism of Gender Identity from Conception to Maturity* (Baltimore: Johns Hopkins University Press, 1972).

Money, John, J. G. Hampson, and J. L. Hampson. "Imprinting and the Establishment of Gender Role." *Archives of Neurology and Psychiatry* 77 (1957): 333–36.

Morgan, Robyn, and Gloria Steinem. "The International Crime of Genital Mutilation." *Ms.* (March 1980): 65–67.

Morland, Iain. "Postmodern Intersex." In *Ethics and Intersex*, edited by Sharon E. Sytsma, 319–31 (Netherlands: Springer, 2006).

———. "Plastic Man: Intersex, Humanism, and the Reimer Case." *Subject Matters: A Journal of Communications and the Self* 3, no. 2/4, no. 1 (2007): 81–98.

———. "Between Critique and Reform: Ways of Reading the Intersex Controversy." In *Critical Intersex*, edited by Morgan Holmes, 191–213 (Farnham: Ashgate, 2009).

———, ed. "Intersex and After" (spec. issue). *GLQ* 15, no. 2 (2009).

———. "Introduction: Lessons from the Octopus." *GLQ* 15, no. 2 (2009): 191–97.

———. "Intersex Treatment and the Promise of Trauma." In *Gender and the Science of Difference: Cultural Politics of Contemporary Science and Medicine*, edited by Jill Fisher, 147–63 (New Brunswick: Rutgers University Press, 2011).

———. "The Injured World: Intersex and the Phenomenology of Feeling." *differences* 23, no. 2 (2012): 20–41.

———. "Cybernetic Sexology." In *Fuckology: Critical Essays on John Money's Diagnostic Concepts*, edited by Lisa Downing, Iain Morland, and Nikki Sullivan, 101–32 (Chicago: University of Chicago Press, 2015).

———. "Gender, Genitals, and the Meaning of Being Human." In *Fuckology: Critical Essays on John Money's Diagnostic Concepts*, edited by Lisa Downing, Iain Morland, and Nikki Sullivan, 69–98 (Chicago: University of Chicago Press, 2015).

Munoz, Jose Esteban. *Disidentifications: Queers of Color and the Performance of Politics* (Minneapolis: University of Minnesota Press, 1999).

Munro, Brenna. "Caster Semenya: Gods and Monsters." *Safundi* 11, no. 4 (2010): 383–96.

Nagar, Richa, and Amanda Lock Swarr. "Introduction." In *Critical Transnational Feminist Praxis*, edited by Amanda Lock Swarr and Richa Nagar, 1–22 (New York: State University of New York Press, 2010).

Nandl, Jacinta. "Germany Got It Right by Offering a Third Gender Option on Birth Certificates." *The Guardian*. November 10, 2013. http://www.theguardian.com/commentisfree/2013/nov/10/germany-third-gender-birth-certificate.

Nash, Jennifer C. "Re-thinking Intersectionality." *Feminist Review* 89 (2008): 1–15.

Naples, Nancy, ed. *The Blackwell Encyclopedia of Gender and Sexuality Studies* (New York: Wiley-Blackwell, 2016).

National Organization for Women (NOW). "NOW Adopts Intersex Resolution." http://www.isna.org/node/170.

Nestle, Joan, Clare Howell, and Riki Wilchins, eds. *GenderQueer: Voices from Beyond the Sexual Binary* (Los Angeles: Alyson Books, 2002).

Njambi, Wairimu Ngaruiya. "Dualisms and Female Bodies in Representations of African Female Circumcision: A Feminist Critique." *Feminist Theory* 5, no. 3 (2004): 281–303.

Noble, Jean Bobby. *Sons of the Movement: FtMs Risking Incoherence on a Post-Queer Cultural Landscape* (Toronto: Women's Press, 2006).

Nussbaum, Martha. "The Professor of Parody." *The New Republic* 220, no. 16 (February 22, 1999): 37–45.

Nyong'o, Tavia. "The Unforgivable Transgression of Being Caster Semenya." *Women and Performance* 20, no. 1 (2010): 95–100.

Oakley, Ann. *Sex, Gender and Society* (London: Maurice Temple Smith, 1972).

Obiora, Leslye. "Bridges and Barricades: Rethinking Polemics and Intransigence in the Campaign Against Female Circumcision." In *Global Critical Race Feminism: An International Reader*, edited by Adrien Katherine Wing, 260–74 (New York: New York University Press, 2000).

One in 2000. Directed by Ajae Clearway. Champaign, IL: Polyvinyl Pictures, 2006.

Ong, Aihwa. *Neoliberalism as Exception: Mutations in Citizenship and Sovereignty* (Durham: Duke University Press, 2006).

Organisation Intersex International (OII). http://www.intersexualite.org/.

———. "FAQ." http://www.intersexualite.org/Organisation_Intersex_International.html.

Organization of American States, "Rights of Lesbian, Gay, Bisexual, Trans, and Intersex Persons." 2013. http://www.oas.org/en/iachr/lgtbi/.

Oyama, Susan, Paul E. Griffiths, and Russell D. Grey, eds. *Cycles of Contingency: Developmental Systems and Evolution* (Massachusetts: MIT Press, 2003).

Parens, Eric, ed. *Surgically Shaping Children: Technology, Ethics, and the Pursuit of Normality* (Baltimore: Johns Hopkins University Press, 2006).

Pateman, Carole. *The Sexual Contract* (Stanford: Stanford University Press, 1988).

Pérez, Emma. *The Decolonial Imaginary: Writing Chicanas into History* (Bloomington: Indiana University Press, 1999).

Preves, Sharon. "Sexing the Intersexed: An Analysis of Sociocultural Responses to Intersexuality." *Signs* 27, no. 2 (2002): 523–56.

———. *Intersex and Identity: The Contested Self* (New Brunswick: Rutgers University Press, 2003).

Prosser, Jay. *Second Skins* (New York: Columbia University Press, 1998).

Puar, Jasbir K. *Terrorist Assemblages: Homonationalism in Queer Times* (Durham: Duke University Press, 2007).

Quijano, Anibal. "Coloniality of Power, Eurocentrism, and Latin America." *Nepantla: Views from South* 1, no. 3 (2000): 533–80.

Rabate, Jean-Michel. *The Future of Theory* (London: Blackwell, 2002).

Reddy, Chandan. *Freedom with Violence: Race, Sexuality, and the US State* (Durham: Duke University Press, 2011).

Reis, Elisabeth. "Divergence or Disorder? The Politics of Naming Intersex." *Perspectives in Biology and Medicine* 50, no. 4 (2007): 535–43.

———. *Bodies in Doubt: An American History of Intersex* (Baltimore: Johns Hopkins University Press, 2012).

Riley, Denise. *Am I That Name? Feminism and the Category of Women in History* (Minneapolis: University of Minnesota Press, 1988).

Roberts, Dorothy. *Fatal Invention: How Science, Politics, and Big Business Re-create Race in the Twenty-First Century* (New York: The New Press, 2011).

Robertson, Claire C. "Getting Beyond the Ew! Factor: Rethinking U.S. Approaches to African Genital Cutting." In *Genital Cutting and Transnational Sisterhood:*

Disputing U.S. Polemics, edited by Stanlie M. James and Clair C. Robertson, 54–86 (Urbana: University of Illinois Press, 2002).

Rogers, Lesley, and Joan Walsh. "Shortcomings of the Psychomedical Research of John Money and Co-Workers into Sex Differences in Behavior: Social and Political Implications." *Sex Roles* 8 (1982): 269–81.

Rosario, Vernon. "Book Review: *Changing Sex: Transsexualism, Technology, and the Idea of Gender*." *Configurations* 4, no. 2 (1996): 243–46.

———. "The Biology of Gender and the Construction of Sex?" *GLQ* 10, no. 2 (2004): 280–87.

———. "An Interview with Cheryl Chase." *Journal of Gay and Lesbian Psychotherapy* 10, no. 2 (2006): 93–104.

———. "The History of Aphallia and the Intersexual Challenge to Sex/Gender." In *A Companion to Lesbian, Gay, Bisexual, Transgender, and Queer Studies*, edited by George E. Haggerty and Molly McGarry, 262–81 (London: Blackwell, 2007).

———. "Quantum Sex: Intersex and the Molecular Deconstruction of Sex." *GLQ* 15, no. 2 (2009): 267–64.

Rosenberg, Jordana. "Butler's 'Lesbian Phallus'; or, What Can Deconstruction Feel?" *GLQ* 9, no. 3 (2003): 393–414.

Roy, Deboleena. "Asking Different Questions: Feminist Practices for the Natural Sciences." *Hypatia* 23, no. 4 (2008): 134–57.

Rubin, David A. "Biochemistry and Physiology of Sex and Gender." In *Wiley-Blackwell Encyclopedia of Gender and Sexuality Studies*, edited by Nancy Naples, 1–7 (New York: Wiley-Blackwell, 2016).

———. " 'An Unnamed Blank that Craved a Name': A Genealogy of Intersex as Gender." *Signs* 37, no. 4 (2012): 883–908.

———. "Women's Studies, Neoliberalism, and the Paradox of the 'Political.' " In *Women's Studies for the Future: Foundations, Interrogations, Politics*, edited by Elizabeth Kennedy and Agatha Beins, 245–61 (New Brunswick: Rutgers University Press, 2005).

Rubin, Gayle. "The Traffic in Women: Notes on the Political Economy of Sex." In *Toward an Anthropology of Women*, edited by Rayna R. Reiter, 157–210 (Boston: Monthly Review Press, 1975).

Salamon, Gayle. "Transfeminism and the Future of Gender." In *Women's Studies on the Edge*, edited by Joan Wallach Scott, 115–38 (Durham: Duke University Press, 2008).

Samuels, Ellen Jean. "Critical Divides: Judith Butler's Body Theory and the Question of Disability." *NWSA Journal* 14, no. 3 (2002): 58–76.

Sánchez, Patricia González, Catalina Velásquez Acevedo, and Sandra Patricia Duque Quintero. "Problemática Jurídica de Los Estados Intersexuales: El Caso Colombiano." ["Legal Problems of Intersex States: The Colombian Case."] *Iatriea* 23, no. 3 (2010): 204–11.

San Francisco Human Rights Commission. http://www.sfgov.org/site/sfhumanrights_index.asp?id=4579.

Sandoval, Chela. *Methodology of the Oppressed* (Minneapolis: University of Minnesota Press, 2000).

Saussure, Ferdinand de. *Course in General Linguistics*. Translated by Roy Harris (Illinois: Open Court Publishing, 1998).

Scott, Joan W. "Experience." In *The Lesbian and Gay Studies Reader*, edited by Henry Abelove, Michele, Aina Barale, and David M. Halperin, 397–415 (New York: Routledge, 1993).

———, ed. *Women's Studies on the Edge* (Durham: Duke University Press, 2008).

Scully, Pamela. "The Trials of Caster Semenya. *The Defenders Online*. July 16, 2010. http://www.thedefendersonline.com/2010/07/16/the-trials-of-caster-semenya/.

Sedgwick, Eve Kosofsky. *Between Men: English Literature and Male Homosocial Desire* (New York: Columbia University Press, 1985).

———. *Epistemology of the Closet* (Berkeley: University of California Press, 1991).

———. "Queer and Now." In *Tendencies*, 1–22 (Durham: Duke University Press, 1994).

———. *Tendencies* (Durham: Duke University Press, 1994).

Semenya, Caster. "Caster Semenya's Comeback Statement in Full." *The Guardian*. March 30, 2010. https://www.theguardian.com/sport/2010/mar/30/caster-semenya-comeback-statement.

Smith, Ben. "Caster Semenya: 'What I Dream of Is to Become Olympic Champion.'" *BBC Sport/Athletics*. May 19, 2015. http://www.bbc.com/sport/0/athletics/32805695.

Smith, David. "Caster Semenya Row: 'Who Are White People to Question the Makeup of an African Girl? It Is Racism.'" *Guardian*. August 22, 2009. http://www.theguardian.com/sport/2009/aug/23/caster-semenya-athletics-gender.

Smith, Bonnie G., and Beth Hutchison, eds. *Gendering Disability* (New Brunswick: Rutgers University Press, 2004).

Soto, Sandra K. *Reading Chican@ Like a Queer: The Demastery of Desire* (Austin: University of Texas Press, 2011).

Spade, Dean. *Normal Life: Administrative Violence, Critical Trans Politics, and the Limits of the Law* (Boston: South End Press, 2011).

Spivak, Gayatri Chakravorty. *In Other Worlds: Essays in Cultural Politics* (New York: Routledge, 1987).

———. "Can the Subaltern Speak?" In *Marxism and the Interpretation of Culture*, edited by Cary Nelson and Laurence Grossberg, 271–314 (Urbana: University of Illinois Press, 1988).

———. *The Postcolonial Critic: Interviews, Strategies, Dialogues*, edited by Sarah Harasym (New York: Routledge, 1990).

———. *Outside in the Teaching Machine* (New York: Routledge, 1993).

———. *A Critique of Postcolonial Reason: Toward a History of the Vanishing Present* (Cambridge: Harvard University Press, 1999).

Stoler, Ann Laura. *Race and the Education of Desire: Foucault's History of Sexuality and the Colonial Order of Things* (Durham: Duke University Press, 1995).

Stoller, Robert. *Sex and Gender: On the Development of Masculinity and Femininity* (New York: Science House, 1968).

Stone, Sandy. "The Empire Strikes Back: A Posttransexual Manifesto." In *The Transgender Studies Reader*, edited by Susan Stryker and Stephen Whittle, 221–35 (New York: Routledge, 2006).

Stryker, Susan. "(De)Subjugated Knowledges: An Introduction to Transgender Studies." In *The Transgender Studies Reader*, edited by Susan Stryker and Stephen Whittle, 1–18 (New York: Routledge, 2006).

———. *Transgender History* (London: Seal Press, 2008).

Stryker, Susan, and Paisley Currah. "General Editors Introduction." *Transgender Studies Quarterly* 1, no. 3 (2014): 303–7.

Stryker, Susan, and Stephen Whittle, eds. *The Transgender Studies Reader* (New York: Routledge, 2006).

Subramaniam, Banu. *Ghost Stories for Darwin: The Science of Variation and the Politics of Diversity* (Urbana: University of Illinois Press, 2014).

Sullivan, Nikki. "'The Price to Pay for Our Common Good': Genital Modification and the Somatechnologies of Cultural (In)Difference." *Social Semiotics* 17, no. 3 (2007): 395–409.

———. "The Matter of Gender." In *Fuckology: Critical Essays on John Money's Diagnostic Concepts*, edited by Lisa Downing, Iain Morland, and Nikki Sullivan, 19–40 (Chicago: University of Chicago Press, 2015).

Swarr, Amanda Lock. *Sex in Transition: Remaking Gender and Race in South Africa* (Albany: State University of New York Press, 2012).

Sytsma, Sharon E., ed. *Ethics and Intersex* (Netherlands: Springer, 2006).

Tamar-Mattis, Anne. "Exceptions to the Rule: Curing the Law's Failure to Protect Intersex Infants." *Berkeley Journal of Gender, Law and Justice* 21 (2006): 59–110.

Tamar-Mattis, Anne, and Milton Diamond. "Managing Variations in Sex Development." *Journal of Pediatric Endocrinology and Metabolism* 20, no. 4 (2007): 552–53.

Tamassia, Arrigo. "Sull'inversione dell'instinto sessuale," *Rivista sperimentale di freniatria e di medicina legale* 4 (1878): 97–119.

Thangaraj, Stanley I. *Desi Hoop Dreams: Pickup Basketball and the Making of Asian-American Masculinity* (New York: New York University Press, 2015).

Vaid, Urvashi. *Virtual Equality: The Mainstreaming of Lesbian and Gay Liberation* (New York: Anchor Books, 1996).

Valentine, David. *Imagining Transgender: An Ethnography of a Category* (Durham: Duke University Press, 2007).

Valentine, David, and Riki Anne Wilchins. "One Percent on the Burn Chart: Gender, Genitals, and Hermaphrodites with Attitude." *Social Text* 15, nos. 3–4 (1997): 215–22.

V-Day. "V-Day Endorses ISNA's Mission to End Violence against Intersex People." http://www.vday.org/node/1497.html#.WOZWwqIrJTZ.

Volcano, Del LaGrace. "The Herm Portfolio." *GLQ* 15, no. 2 (2009): 261–65.

Walker, Alice. *Possessing the Secret of Joy* (New York: Pocket Books, 1992).

———. *The Trouble with Normal* (New York: The Free Press, 1999).

Webster, Fiona. "The Politics of Sex and Gender: Benhabib and Butler Debate Subjectivity." *Hypatia* 15, no. 1 (2000): 1–22.

Wiegman, Robin. "Feminism, Institutionalism, and the Idiom of Failure." *differences* 11, no. 3 (1999): 107–36.

———."What Ails Feminist Criticism? A Second Opinion." *Critical Inquiry* 25, no. 2 (1999): 362–79.

———. "Academic Feminism Against Itself." *NWSA Journal* 14, no. 2 (2002): 18–34.

———. "The Progress of Gender: Whither 'Women'?" In *Women's Studies on Its Own*, edited by Robyn Wiegman, 106–40 (Durham: Duke University Press, 2002).

———, ed. *Women's Studies on Its Own* (Durham: Duke University Press, 2002), 106–40.

———. "The Desire for Gender." In *The Blackwell Companion to Lesbian, Gay, Bisexual, Transgender, and Queer Studies*, edited by George E. Haggerty and Molly McGarry, 217–36 (London: Blackwell, 2007).

———. *Object Lessons* (Durham: Duke University Press, 2012).

Willey, Angela. *Undoing Monogamy: The Politics of Science and the Possibilities of Biology* (Durham: Duke University Press, 2016).

Wilson, Elizabeth A. *Gut Feminism* (Durham: Duke University Press, 2015).

———. *Neural Geographies: Feminism and the Microstructure of Cognition* (New York: Routledge, 1998).

———. *Psychosomatic: Feminism and the Neurological Body* (Durham: Duke University Press, 2004).

Wittig, Monique. "One Is Not Born a Woman." In *The Lesbian and Gay Studies Reader*, edited by Henry Abelove, Michele Aina Barale, and David M. Halperin, 103–9 (New York: Routledge, 1993).

Wolf, Naomi. *The Beauty Myth: How Images of Beauty Are Used against Women* (New York: Harper, 2004).

Wolfe, Cary. *What Is Posthumanism?* (Minneapolis: University of Minnesota Press, 2009).

———, ed. *Zoontologies* (Minneapolis: University of Minnesota Press, 2003).

Young, Iris Marion. *Global Challenges: War, Self-Determination, and Responsibility for Justice* (New York: Polity, 2006).

Zach, Naomi. *Inclusive Feminism: A Third Wave Theory of Women's Commonality* (Lanham, MD: Rowman and Littlefield, 2005).

Zirin, Dave, and Sherry Wolf. "Caster Semenya: The Idiocy of Sex Testing." *The Nation*, August 21, 2009. http://www.thenation.com/doc/20090831/zirin_wolf.

Zuma, Jacob. "IAAF Publicly Humiliated Caster Semenya." *Politics Web*. August 25, 2009. http://www.politicsweb.co.za/politicsweb/view/politicsweb/en/page71639?oid5140761&sn5Detail.

Index